T0201346

BEHAVIORAL MODELING AND PREDISTORTION OF WIDEBAND WIRELESS TRANSMITTERS

BEHAVIORAL MODELING AND PREDISTORTION OF WIDEBAND WIRELESS TRANSMITTERS

Fadhel M. Ghannouchi
Oualid Hammi
Mohamed Helaoui

WILEY

Library of Congress Cataloging-in-Publication Data

Ghannouchi, Fadhel M., 1958-
  Behavioral modelling and predistortion of wideband wireless transmitters / Fadhel Ghannouchi, Oualid Hammi,
Mohamed Helaoui.
       pages cm
  Includes bibliographical references and index.
  ISBN 978-1-118-40627-4 (cloth : alk. paper) 1. Wireless communication systems–Mathematical models.
2. Broadband communication systems–Mathematical models. 3. Signal theory (Telecommunication)–Mathematics.
4. Telecommunication–Transmitters and transmission–Mathematics. 5. Nonlinear systems–Mathematical models.
6. Electric distortion–Mathematical models. I. Title.
  TK5102.83.G46 2015
  621.3841′31—dc23

2015006179

ISBNs
hardback: 978-1118-40627-4
ePDF: 978-1119-00443-1
epub: 978-1119-00444-8
oBook: 978-1119-00442-4

Set in 11/13pt Times by SPi Global, Chennai, India
Printed and bound in Singapore by Markono Print Media Pte Ltd

1 2015

Contents

About the Authors

Dr Fadhel M. Ghannouchi was born in Gabes, Tunisia in 1958. He is currently Professor and iCORE/CRC Chair at the Electrical and Computer Engineering Department of The Schulich School of Engineering of the University of Calgary and founding Director of Intelligent RF Radio Laboratory (www.iradio.ucalgary.ca). His research interests are in the areas of microwave instrumentation and measurements, nonlinear modeling of microwave devices and communications systems, design of power and spectrum efficient microwave amplification systems, and design of intelligent RF transceivers for wireless and satellite communications. His research activities have led to over 600 publications and 15 US patents (three pending) and four books. He is the co-founder of three of the University's spin-off companies. As an educator and mentor, he has supervised and graduated more than 80 Masters and PhD students and has supervised more than 30 post-doctoral fellows. Dr Ghannouchi is an IEEE Fellow, IET Fellow, Fellow of Engineering Institute of Canada (EIC), Fellow of the Canadian Academy of Engineering (CAE), and Fellow of the Royal Society of Canada (RSC). Dr Ghannouchi has also been an IEEE Distinguish Microwave Lecturer (2009–2012).

Dr Oualid Hammi was born in Tunis, Tunisia in 1978. He received a B.Eng. degree from the École Nationale d'Ingénieurs de Tunis, Tunis, Tunisia, in 2001, an MSc degree from the École Polytechnique de Montréal, Montréal, QC, Canada, in 2004, and his PhD from the University of Calgary, Calgary, AB, Canada, in 2008, all in electrical engineering. He is currently an Associate Professor with the Department of Electrical Engineering, King Fahd University of Petroleum and Minerals, Dhahran, Saudi Arabia. He is also an adjunct researcher with the Intelligent RF Radio Laboratory (iRadio Lab), Schulich School of Engineering, University of Calgary. He has authored and co-authored over 80 publications and 7 US patents (six pending), and is a regular reviewer for several IEEE transactions and other journals. His research interests include the characterization, behavioral modeling, and linearization of radiofrequency power amplifiers and transmitters, and the design of energy efficient linear transmitters for wireless communication systems.

Mohamed Helaoui was born in Tunis, Tunisia in 1979. He received his MSc degree in communications and information technology from the École Supérieure des Communications de Tunis, Tunisia, in 2003 and his PhD in electrical engineering from the University of Calgary in 2008. He is currently an Associate Professor at the Department of Electrical and Computer Engineering at the University of Calgary. His current research interests include digital signal processing, power efficiency enhancement for wireless transmitters, switching mode power amplifiers, six-port receivers, and advanced transceiver design for software defined radio and millimeter-wave applications. His research activities have led to over 100 publications and eight patents (three pending).

Preface

Wireless systems are offering a wide variety of services to an ever increasing number of users. Undeniably, this connectivity has contributed to enhancing the quality of life. Though, the proliferation of wireless handheld devices and base stations led to an alarming downside due to their environmental impact. In fact, the carbon footprint of the wireless communication infrastructure is reaching unprecedented levels. This stimulated a global awareness about the need to reduce base stations energy consumption. In order to make communication systems more eco-friendly and "greener", significant research work is being carried out at various aspects of base station design. This includes, among other things, scaling of energy needs depending on the traffic and network load, improving the ratio of quality of service to radiofrequency power, and increasing the overall efficiency of the base station. A closer look at base stations power consumption reflects that their overall efficiency can be significantly improved by increasing that of the radio frequency front end and especially the power amplifier. This would not only make communication systems greener but also reduce their deployment and running costs in terms of capital expenditure (CAPEX) and operational expenditure (OPEX), and result in substantial financial benefits.

Technically, building power amplifiers with peak power efficiencies as high as 80% has become feasible thanks to the development of new transistor technologies and new classes of operation such as switching mode. However, getting such high efficiencies from power amplifiers handling modern wireless communication systems is a tricky challenge. In fact, and due to the nature of the highly varying envelop signals being transmitted, base station power amplification systems have to be highly linear and meet the spectrum emission masks set by standardization and regulatory authorities. This requires the use of linearization techniques, which virtually make the power amplifier linear over its entire power range, thus allowing operation with less power back-off, and hence resulting in higher efficiencies compared to what could have been obtained from the same amplifier if no linearization was adopted. In this context, digital predistortion has received tremendous attention from the industrial and academic communities and incontestably appears to be the preferred technology for base station power amplifier linearization.

Conceptually, behavioral modeling and digital predistortion are intimately related. They are often referred to as forward and reverse modeling, respectively. This book focuses on the behavioral modeling and digital predistortion of wideband power amplifiers and transmitters. It compiles a wide range of topics related to this theme. The book is organized in 10 chapters, which can be organized into three parts. Chapters 1–3 set the ground for the remainder of the book by introducing the key parameters used to model and characterize the nonlinear behavior of wireless transmitters in Chapter 1, classifying and discussing the theory of dynamic nonlinear systems in Chapter 2, and providing a review of model performance evaluations metrics in Chapter 3. The second part of the book, Chapters 4–7, is a thorough review of behavioral models and predistortion functions that encompasses quasi-memoryless models in Chapter 4, memory polynomial based models in Chapter 5, box-oriented models in Chapter 6, and neural networks based models in Chapter 7. These models are introduced and their specificities discussed. The last part of the book, Chapters 8–10, is application oriented and provides comprehensive and insightful information about the use, in an experimental environment, of the models described earlier in the book. Chapter 8 covers the acquisition of the device-under-test (DUT) input and output data and its processing prior to the model identification. Chapter 9 is devoted to baseband digital predistortion and its practical aspects. Chapter 10 concludes the book by exposing recent trends in behavioral modeling and digital predistortion such as joint quadrature impairment compensation and digital predistortion, as well as the predistortion of dual-band and multi-input multi-output (MIMO) transmitters.

The book chapters are complemented with a software tool available through the Wiley website (www.wiley.com/go/Ghannouchi/Behavioral) that implements several of the topics discussed in the book and can be used to demonstrate these topics in a more tangible way.

Acknowledgments

We would like to gratefully acknowledge the help and support received from friends, colleagues, support staff, and students, both past and present at iRadio Laboratory, University of Calgary, Calgary; Poly-GRAMES Research Center, Ecole Polytechnique, Montreal; and King Fahd University of Petroleum and Minerals, KSA. We are grateful to our great students and researchers; this book could not have been completed without their fruitful research. In addition, we would like to thank C. Heys and A. Congreve for their help in proofreading and formatting the manuscript, and Ivana d'Adamo for her administrative support. The authors would also like to thank the IEEE for their acceptance and courtesy in allowing them to reproduce several figures and illustrations published elsewhere in their journals and/or conferences.

Dr O. Hammi acknowledges the support provided by the Deanship of Scientific Research at King Fahd University of Petroleum and Minerals (KFUPM) under research grant IN121039.

Dr F. M. Ghannouchi and M. Helaoui acknowledge the main sponsors and financial supporters of the Intelligent Radio technology Laboratory (iRadio Lab), Alberta Innovates Technology Futures (AITF), Alberta, Canada, the Canada Research Chairs (CRC) program, and Natural Sciences and Engineering Research Council of Canada (NSERC).

Finally, we would like to profoundly thank our respective spouses; Ilhem, Asma, and Imen; and our kids for their understanding and patience throughout the many evenings and weekends taken to prepare this book, as well as our parents for their encouragement and valuable support in our early professional years as graduate students and young researchers.

1

Characterization of Wireless Transmitter Distortions

1.1 Introduction

Wireless transmitters designed for modern communication systems are expected to handle wideband amplitude and phase modulated signals with three major performance metrics: linearity, bandwidth, and power efficiency. First, linearity requires the minimization of distortions mainly caused by the transmitter's radio frequency (RF) analog circuitry in order to preserve the quality of the transmitted signal and avoid any loss of information during the transmission process. Second, bandwidth is critical for multi-carrier and multi-band communication systems. Moreover, wider bandwidths are needed to accommodate higher data rates. Third, power efficiency is an important consideration that affects the deployment and operating costs of communication infrastructure as well as environmental impact.

In general, distortions refer to the alteration of the signal due to the imperfections of the transmitter's hardware. Distortions observed in wireless transmitters have various origins such as frequency response distortions, harmonic distortions, amplitude and phase distortions, and group delay distortions, in addition to modulator impairments (including direct current (DC) offset, gain, and phase imbalance), and so on. Among these distortions, the predominant ones are those due to the nonlinearity present in the transmitter's RF front end and mainly the RF power amplifier (PA). Indeed, wireless transmitters are made of a cascade of several stages including digital-to-analog conversion, modulation, frequency up-conversion, filtering, and amplification as illustrated in Figure 1.1. Among these subsystems, the PA is identified as the major source of nonlinear distortions. Thus, modeling and compensating for the transmitter nonlinear distortions is often trimmed down to the modeling and compensation of the PA's nonlinearity.

Behavioral Modeling and Predistortion of Wideband Wireless Transmitters, First Edition.
Fadhel M. Ghannouchi, Oualid Hammi and Mohamed Helaoui.
© 2015 John Wiley & Sons, Ltd. Published 2015 by John Wiley & Sons, Ltd.

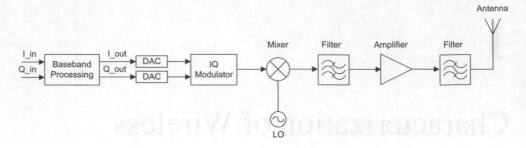

Figure 1.1   Simplified block diagram of a typical wireless transmitter

In the remainder of this chapter, the nonlinearity of RF PAs will be described and major metrics used to quantify nonlinear distortions will be presented.

1.1.1   RF Power Amplifier Nonlinearity

The nonlinearity of the PA depends mainly on its class of operation and topology. Classes of operation include the linear class A, the mildly nonlinear class AB, as well as highly nonlinear classes such as C, D, and E. The topology refers to whether the power amplification system is built using single-ended amplifiers or more advanced architectures such as Doherty, linear amplification using nonlinear components (LINC), envelope tracking, and so on. The design of power amplification systems is always subject to the unavoidable antagonism between linearity and power efficiency [1]. The objective is to design a power amplification system, or more generally, a transmitter that meets the linearity requirements with the highest possible power efficiency. The approach often consists of maximizing the power efficiency of the amplification stage while maintaining its distortions to a reasonable amount that can be compensated for at the system level using linearization techniques such as feedforward or predistortion [2]. Figure 1.2 shows the measured gain and power efficiency of a Gallium Nitride (GaN) based Doherty PA driven by a four-carrier wideband code division multiple access (WCDMA) signal and operating around a carrier frequency of 2140 MHz. This figure clearly illustrates the power efficiency versus linearity dilemma as low power efficiency is observed for low input power levels when the amplifier is operating in its linear region where the gain is constant. Conversely, higher power efficiency is obtained for large input power levels that drive the amplifier into its nonlinear region.

1.1.2   Inter-Modulation Distortion and Spectrum Regrowth

Transmitters' nonlinearity causes the appearance of unwanted frequency components at the output of the transmitter. To better understand the effects of the transmitter's nonlinearity on the transmitted signal, the case of a two-tone signal passing through

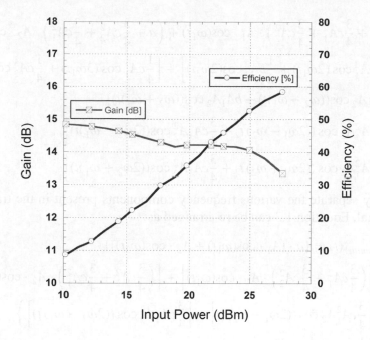

Figure 1.2 Gain and power efficiency characteristics of a power amplifier prototype

a third order memoryless nonlinear systems is considered in the example next. In this case:

- The transmitter's nonlinearity is modeled by a third order polynomial function according to the following equation:

$$x_{out\_Transmitter}(t) = a \cdot x_{in\_Transmitter}(t) + b \cdot x^2_{in\_Transmitter}(t) + c \cdot x^3_{in\_Transmitter}(t) \quad (1.1)$$

where $x_{in\_Transmitter}$ and $x_{out\_Transmitter}$ are the time domain waveforms at the input and the output of the transmitter, respectively. a, b, and c are the model coefficients.
- The input signal $x_{in\_Transmitter}$ is a two-tone signal given by:

$$x_{in\_Transmitter}(t) = A_1 \cdot \cos(\omega_1 t) + A_2 \cdot \cos(\omega_2 t) \quad (1.2)$$

where A_1 and A_2 are the magnitudes of each of the two tones, and ω_1 and ω_2 are their angular frequencies with $\omega_2 > \omega_1$.

By combining Equations 1.1 and 1.2, the transmitter's output for the two-tone input signal can be expressed as:

$$x_{out\_Transmitter}(t)$$
$$= \left[\frac{1}{2}bA_1^2 + \frac{1}{2}bA_2^2 \right]$$

$$+\left[\left(a+\frac{3}{4}cA_1^2+\frac{3}{4}cA_2^2\right)\cdot A_1\cdot\cos(\omega_1 t)+\left(a+\frac{3}{4}cA_2^2+\frac{3}{4}cA_1^2\right)\cdot A_2\cdot\cos(\omega_2 t)\right]$$

$$+\left[\frac{1}{2}bA_1^2\cos(2\omega_1 t)+\frac{1}{2}bA_2^2\cos(2\omega_2 t)\right]+\left[\frac{1}{4}cA_1^3\cos(3\omega_1 t)+\frac{1}{4}cA_2^3\cos(3\omega_2 t)\right]$$

$$+[bA_1A_2\cos((\omega_2-\omega_1)t)+bA_1A_2\cos((\omega_2+\omega_1)t)]$$

$$+\left[\frac{3}{4}cA_1^2A_2\cos((2\omega_1-\omega_2)t)+\frac{3}{4}cA_1A_2^2\cos((2\omega_2-\omega_1)t)\right]$$

$$+\left[\frac{3}{4}cA_1^2A_2\cos((2\omega_1+\omega_2)t)+\frac{3}{4}cA_1A_2^2\cos((2\omega_2+\omega_1)t)\right] \tag{1.3}$$

To clearly separate the various frequency components present in the transmitter's output signal, Equation 1.3 can be re-arranged as:

$$x_{out\_Transmitter}(t)=\{a\cdot[A_1\cdot\cos(\omega_1 t)+A_2\cdot\cos(\omega_2 t)]\}$$

$$+\left\{\left[\left(\frac{3}{4}cA_1^2+\frac{3}{4}cA_2^2\right)\cdot A_1\cdot\cos(\omega_1 t)\right]+\left[\left(\frac{3}{4}cA_2^2+\frac{3}{4}cA_1^2\right)\cdot A_2\cdot\cos(\omega_2 t)\right]\right\}$$

$$+\left\{\left[\frac{3}{4}cA_1^2A_2\cos((2\omega_1-\omega_2)t)\right]+\left[\frac{3}{4}cA_1A_2^2\cos((2\omega_2-\omega_1)t)\right]\right\}$$

$$+\left\{\left[\frac{1}{2}bA_1^2+\frac{1}{2}bA_2^2\right]+[bA_1A_2\cos((\omega_2-\omega_1)t)]\right\}$$

$$+\left\{\frac{1}{2}b\left[A_1^2\cos(2\omega_1 t)+A_2^2\cos(2\omega_2 t)\right]+[bA_1A_2\cos((\omega_2+\omega_1)t)]\right\}$$

$$+\left\{\frac{1}{4}c\left[A_1^3\cos(3\omega_1 t)+A_2^3\cos(3\omega_2 t)\right]\right.$$

$$\left.+\frac{3}{4}cA_1A_2\left[A_1\cos((2\omega_1+\omega_2)t)+A_2\cos((2\omega_2+\omega_1)t)\right]\right\} \tag{1.4}$$

In this latter equation, the term between the first brackets ({}) in the right hand side represents the linearly amplified version of the input signal, while the second term corresponds to the distortions introduced by the transmitter's nonlinearity at the fundamental frequencies (these are the same as the input signal's frequencies). The remaining terms describe the mixing and harmonic frequency products that either fall in the close vicinity of the useful signal and thus cannot be removed by filtering, or are away from the useful signal (around DC or the harmonics). The latter are less critical as they can be removed by filtering the transmitter's output signal. The frequency domain representation of the transmitter's input and output signals given by Equations 1.2 and 1.4 are illustrated in Figure 1.3.

The frequency components present at the output of the nonlinear transmitter driven by a two-tone input signal are summarized in Table 1.1. These can be categorized in three groups:

- *The useful signal:* comprised of the linearly amplified fundamental frequency components.

- *The unwanted signals that can be removed by filtering:* these include the DC components, the second and third order harmonics, second order inter-modulation distortions, as well as out-of-band third order inter-modulation distortions.
- *The unwanted signals that cannot be filtered:* this includes the distortions that appear at the same frequencies as the input signal, and in-band third order inter-modulation products that are too close to the fundamental components to be filtered. For higher order nonlinear systems, additional even order in-band inter-modulation products are observed in the close vicinity of the useful signal.

The analysis presented here can be generalized to an *N*th order nonlinear model of the transmitter. In such case, up to the *N*th order harmonics and *N*th order mixing products will be generated at the output of the nonlinear transmitter [3, 4].

The study of PA and transmitter nonlinearities using two-tone and multi-tone signals is commonly used for understanding the origins of inter-modulation distortions for signals having discrete frequency spectrum components and can be used to derive closed form expressions of these distortions under two-tone or multi-tone input signals [5, 6]. Such results can be extrapolated to predict the behavior of the nonlinear system when driven by communications and broadcasting signals having characteristics comparable to that of synthetic multi-tone signals. However, when practical communication signals are used, the input signal's spectrum is continuous

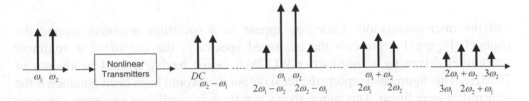

Figure 1.3 Frequency domain output of a nonlinear transmitter driven by a two-tone signal

Table 1.1 Frequency components at the output of a nonlinear transmitter for a two-tone input signal

Angular frequency	Designation
0	DC components
ω_1 and ω_2	Fundamental
$2\omega_1$ and $2\omega_2$	Second harmonics
$3\omega_1$ and $3\omega_2$	Third harmonics
$\omega_2 - \omega_1$ and $\omega_2 + \omega_1$	Second order inter-modulation products
$2\omega_1 - \omega_2$ and $2\omega_2 - \omega_1$	In-band third order inter-modulation products
$2\omega_1 + \omega_2$ and $2\omega_2 + \omega_1$	Out-of-band third order inter-modulation products

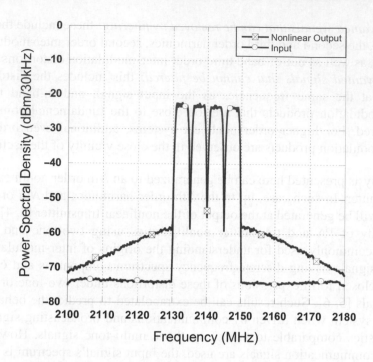

Figure 1.4   Output spectrum of a nonlinear transmitter driven by a multi-carrier WCDMA signal

and the inter-modulation distortions appear as a spectrum regrowth around the channel. Figure 1.4 presents the measured spectra at the output of a nonlinear transmitter driven by a four-carrier WCDMA signal having a total bandwidth of 20 MHz. This figure also reports the ideal output that would have been obtained if the transmitter were linear. This figure shows that there is significant spectrum regrowth that will create interferences with the adjacent channels. Such a transmitter does not meet the spectrum emission mask of the WCDMA standard and unavoidably requires linearization.

1.2   Impact of Distortions on Transmitter Performances

The nonlinearity of the PA depends on the input power level or equivalently on the input signal's amplitude. Thus, phase modulated signals having constant envelopes are not affected by the nonlinearity of the PA. Conversely, amplitude modulated signals are distorted by the nonlinearities. Almost all modern communication and broadcasting systems employ compact complex modulation schemes such as high order quadrature amplitude modulations (16QAM, 64QAM, etc.) and advanced multiplexing techniques, for example, orthogonal frequency division multiplexing (OFDM), and code division multiple access (CDMA), which result in amplitude modulated signals having strong envelope fluctuations. These signals are characterized by their peak-to-average power ratio (PAPR) that is given by:

$$PAPR_{dB} = 10 \times \log_{10}\left(\frac{P_{max,W}}{P_{avg,W}}\right) = P_{max,dBm} - P_{avg,dBm} \qquad (1.5)$$

where $PAPR_{dB}$ is the signal's PAPR expressed in dB. $P_{max,W}$ and $P_{avg,W}$ are the signal's maximum and average power levels expressed in watts, respectively. Similarly, $P_{max,dBm}$ and $P_{avg,dBm}$ are the signal's maximum and average power levels expressed in dBm, respectively.

Typical PAPR values for modern communication systems are in the range of 10–13 dB. These can be reduced by several decibels using crest factor reduction (CFR) techniques [7–9]. The PAPR of the signal and its probability distribution functions are critical parameters that need to be considered when dealing with amplifier and transmitter nonlinearities. Indeed, to linearly amplify high PAPR signals without linearizing the amplifier, one must make sure that the maximum peak power of the input signal to be amplified remains within the linear region of the PA. This will impact the power efficiency of the system. To illustrate this concept of brute force linear amplification graphically, the gain and drain efficiency characteristics of a commercial PA are presented in Figure 1.5. In this figure, the gain and power efficiency are reported as a function of the output power back-off (OPBO) that is defined as:

$$OPBO_{dB} = P_{out,dBm} - P_{out,sat,dBm} \qquad (1.6)$$

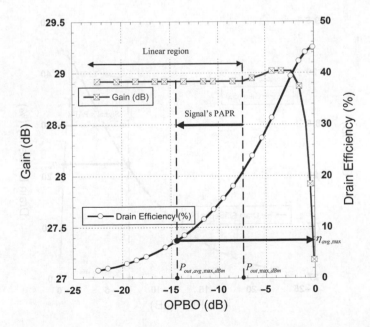

Figure 1.5 Gain and efficiency considerations in brute force linear amplification

where $OPBO_{dB}$ is the OPBO expressed in dB. $P_{out,dBm}$ and $P_{out,sat,dBm}$ refer to the dBm values of the amplifier's operating output power and the amplifier's output power at saturation, respectively.

According to the results of Figure 1.5, to ensure a linear behavior of the considered amplifier, the maximum peak output power ($P_{out,max,dBm}$) should not exceed −7 dB OPBO. Thus, the maximum average output power ($P_{out,avg,max,dBm}$) will be:

$$P_{out,avg,max,dBm} = P_{out,max,dBm} - PAPR_{dB} \qquad (1.7)$$

In Equation 1.7, $PAPR_{dB}$ is the signal's PAPR expressed in dB. In this example, the signal's PAPR is assumed to be 7 dB.

Given this restriction on the maximum operating average power of the amplifier, the maximum average drain efficiency ($\eta_{avg,max}$) of the brute force linear amplifier will be less than 10%. This noticeably low power efficiency represents the maximum efficiency achievable from this amplifier if operated without a linearization technique. Conversely, if the same amplifier is used in conjunction with a linearization technique, for example, using digital predistortion (DPD), it will be able to operate linearly over its full output power range up to saturation. As graphically illustrated in Figure 1.6, the maximum average output power of the amplifier will be higher, which enables increased power efficiency. This example shows that by using linearization techniques, the maximum efficiency of the amplifier can be raised from 8 to 23%, which represents a substantial gain in power efficiency. It is worth mentioning that the amplifier used in this graphical analysis is optimized for linearity. Though, if a

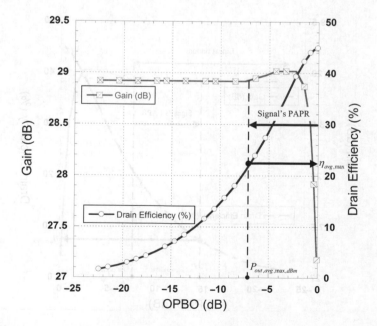

Figure 1.6 Gain and efficiency considerations in linearized power amplifiers

power efficient amplifier prototype is considered, a more important efficiency gain can be obtained with operating efficiencies of the PA in the range of 50%.

This brief discussion clearly shows the impact of distortions on the system efficiency as they constrain the brute force amplifier to work with large back-off levels to guarantee linear amplification. It also highlights the significant power efficiency improvement that can be obtained by using a DPD technique.

The cascade of the PA and the digital predistorter will behave as a linear amplification system whose gain can be set by controlling the small signal gain of the predistorter. It is a common misconception to think that the choice of the small signal gain of the predistortion and thus the gain of the linearized amplifier will influence the power efficiency performance of the linearized amplifier. Indeed, when seen as a function of the output power, the drain efficiency of the linearized amplifier will remain quasi unchanged [10]. The impact of the gain normalization on the DPD performance will be thoroughly discussed in Chapter 9.

These efficiency figures do not take into account the energy consumption of the linearization circuitry. Typical DPD circuitry has a power consumption in the range of a few watts. This power consumption needs to be taken into consideration when calculating the overall efficiency of the linearized amplifier. Obviously, the use of the predistortion technique for efficiency/linearity trade-off enhancement is a viable solution only when the predistorter's power consumption does not compromise the overall efficiency of the linearized amplifiers. Accordingly, and as rule of thumb, DPD is practically employed for PAs with output power that exceeds 10 W: Though accurate calculations can be made to decide on the suitability of DPD to improve the system performance compared to the case of a brute force amplifier topology based on the amplifier's power capability, its efficiency, and the predistorter's power consumption.

1.3 Output Power versus Input Power Characteristic

The output power versus input power (P_{out} vs. P_{in}) characteristic is commonly used to characterize the transfer function of amplifiers. This characteristic relates the input power of the device under test (DUT) at the fundamental frequency to its output power at the same frequency. When both power levels are expressed in watts, the slope of the P_{out} vs. P_{in} characteristic represents the linear gain of the system. Most commonly, the power levels are expressed in dBm. In such case, the slope of the P_{out} vs. P_{in} characteristic is equal to unity and the gain in dB corresponds to the y-intercept point (i.e., the value of the output power for a 0 dB m input power).

In the absence of memory effects, the P_{out} vs. P_{in} characteristic appears as a one to one mapping function that increases linearly with the input power. As the amplifier is driven into its nonlinear region, a gain compression appears as the actual output power becomes lower than the linearly amplified version of the input power. The amount of compression introduced by the amplifier increases until it reaches the saturation power. Figure 1.7 presents a sample P_{out} vs. P_{in} characteristic of an amplifier

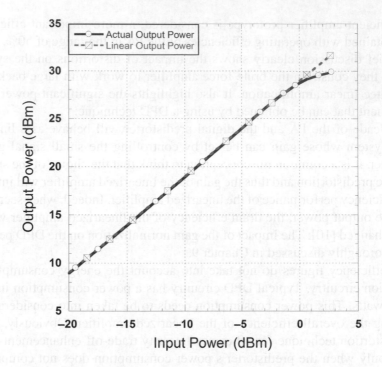

Figure 1.7 Sample output power versus input power characteristic

optimized for linearity. One can observe that the gain compression of the amplifier becomes noticeable only a few dBs before the saturation for input power levels beyond 0 dB m. With amplifiers optimized for efficiency, the gain compression is observed over a wider input power range starting from as early as 10 dB below the maximum input power.

The P_{out} vs. P_{in} characteristic is straightforward to derive as it only requires scalar measurements both at the input and output of the DUT. This can be performed using a network analyzer or a set up that comprises a signal generation instrument and a power measurement instrument such as a power meter or a spectrum analyzer. The P_{out} vs. P_{in} characteristic can be measured under a wide range of drive signals such as continuous wave (CW), multi-tone, or modulated signals.

1.4   AM/AM and AM/PM Characteristics

The P_{out} vs. P_{in} characteristic is a basic and incomplete means of characterizing non-linear transmitters and PAs driven by modulated signals. Indeed, a more comprehensive representation that includes amplitude as well as phase information is needed. In the most general case, a dynamic nonlinear transmitter is fully described by a set of four characteristics, namely the amplitude modulation to amplitude modulation (AM/AM) characteristic, the amplitude modulation to phase modulation (AM/PM) characteristic, the phase modulation to phase modulation (PM/PM) characteristic, and the phase modulation to amplitude modulation (PM/AM) characteristic. PA distortions are amplitude dependant and phase modulated signals (having constant amplitudes) are not affected by the PA distortions. Thus, PAs are mainly characterized by their AM/AM and AM/PM characteristics. Conversely, transmitters might exhibit PM/AM and PM/PM distortions that are mainly due to the gain and phase imbalances in the frequency up-conversion stage and/or when the transmitter has a non-flat frequency response over a bandwidth equal to that of the input signal. Contrary to the AM/AM and AM/PM distortions generated by the unavoidably nonlinear behavior of the PA, the PM/AM and PM/PM distortions can be minimized by a careful design of the transmitter. So far, these have often been considered to have an insignificant impact on the performance of a behavioral model or a digital predistorter. With the adoption of multi-carriers and multi-band power amplification systems where the bandwidth of the signal to be transmitted is large enough to observe on a non-flat frequency response of the PA, the contribution of the PM/AM and PM/PM is becoming more significant and their inclusion in next generation behavioral models and predistorters is becoming inevitable.

Let's consider a DUT driven by a modulated input signal. x_{in} and x_{out} refer to the baseband complex waveforms corresponding to the DUT's input and output signals, respectively. The in-phase and quadrature components of the signals x_{in} and x_{out} are defined as:

$$\begin{cases} x_{in} = I_{in} + jQ_{in} \\ x_{out} = I_{out} + jQ_{out} \end{cases} \tag{1.8}$$

Under the assumption that this DUT, to be modeled or equivalently linearized, does not exhibit PM/AM and PM/PM distortions, its instantaneous complex gain, G, is solely a function of the input signal's magnitude and is given by:

$$G(|x_{in}|) = |G(|x_{in}|)| \cdot \overline{|G(|x_{in}|)}| \tag{1.9}$$

where $|G(|x_{in}|)|$ and $\overline{|G(|x_{in}|)}|$ represent the magnitude and phase of the instantaneous complex gain $G(|x_{in}|)$, respectively; and are expressed as a function of the input and output complex baseband waveforms according to:

$$|G(|x_{in}|)| = \frac{|x_{out}|^2}{|x_{in}|^2} = \frac{I_{out}^2 + Q_{out}^2}{I_{in}^2 + Q_{in}^2} \tag{1.10}$$

$$\overline{|G(|x_{in}|)} = \underline{|x_{out}} - \underline{|x_{in}} = \tan^{-1}\left(\frac{Q_{out}}{I_{out}}\right) - \tan^{-1}\left(\frac{Q_{in}}{I_{in}}\right) \tag{1.11}$$

The AM/AM characteristic of the DUT is obtained by plotting the magnitude of its instantaneous gain ($|G(|x_{in}|)|$), typically expressed in dB, as a function of the DUT's instantaneous input power. It is also possible, though less conventional to report the AM/AM characteristic as function of the DUT's output power. Similarly, the AM/PM

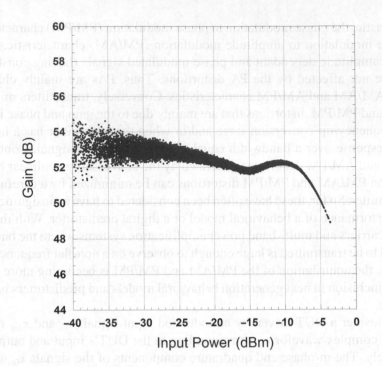

Figure 1.8 Sample AM/AM characteristic of a power amplifier

characteristic of the DUT is the one that reports the phase of the instantaneous gain $(\angle G(|x_{in}|))$, usually expressed in degrees, as a function of the DUT's input or output power. Sample AM/AM and AM/PM characteristics are reported in Figures 1.8 and 1.9, respectively. These figures provide insightful information about the nonlinear behavior of the DUT. In fact, the shape of the AM/AM and AM/PM characteristics provide information about how severe the nonlinearity of the DUT is. Similarly, the dispersion of these two characteristics is a qualitative indication about the memory effects of the device.

1.5   1 dB Compression Point

The 1 dB compression point is a figure of merit commonly used to characterize the power capabilities of PAs along with their linearity. In the P_{out} vs. P_{in} characteristic, the 1 dB compression point is the one for which the actual output power of the amplifier is 1 dB lower than what it would have been if the amplifier was linear (and having a gain equal to its small signal gain). This definition is illustrated graphically in Figure 1.10, which reports the P_{out} vs. P_{in} characteristics of the actual and ideal amplifier. The ideal amplifier characteristic represents the extrapolated version of the linear portion of the actual amplifier's P_{out} vs. P_{in} characteristic. This figure shows that the 1 dB compression point can be defined either with respect to the input power ($P_{1dB,in}$ in Figure 1.10) or with reference to the output power ($P_{1dB,out}$ in Figure 1.10).

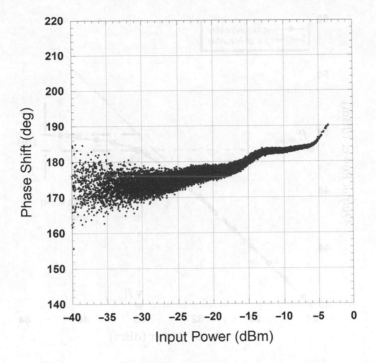

Figure 1.9 Sample AM/PM characteristic of a power amplifier

Though, the 1 dB compression point is commonly reported with respect to the output power of the device.

Similarly, the 1 dB compression point can be defined from the AM/AM character-istic. In this case, it corresponds to the power level for which the gain of the amplifier is 1 dB lower than its small signal linear value. From the AM/AM characteristic, the $P_{1dB,in}$ can be graphically determined as illustrated in Figure 1.11. If the small signal gain of the amplifier is denoted as G_{SS}, then the 1 dB compression point output power ($P_{1dB,out}$) can be obtained according to:

$$P_{1dB,out} = P_{1dB,in} + (G_{SS} - 1) \tag{1.12}$$

For a given DUT, the 1 dB compression point can vary depending on the test signal (CW versus modulated signals). The 1 dB compression point concept can be extended to the X-dB compression point. The X-dB compression point is defined in a way sim-ilar to that of the 1-dB compression point but for a gain compression of X-dB rather than 1 dB. Thus, the 3-dB compression point is the point of the P_{out} vs. P_{in} charac-teristic for which the actual output power of the amplifier is 3 dB less than what it would have been if the amplifier was linear; it is also the point of the AM/AM char-acteristic for which the gain of the device is 3 dB lower than its small signal value. The X-dB compression point can be used for the system level design of power ampli-fication stages as well as building equation-based behavioral models in simulation software.

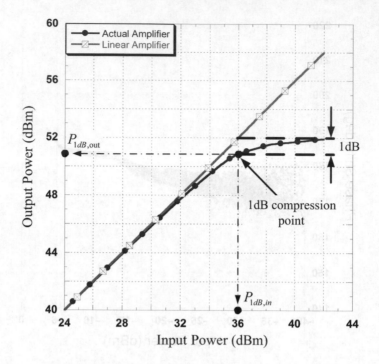

Figure 1.10   Graphical definition of the 1 dB compression point from P_{out} vs. P_{in} characteristic

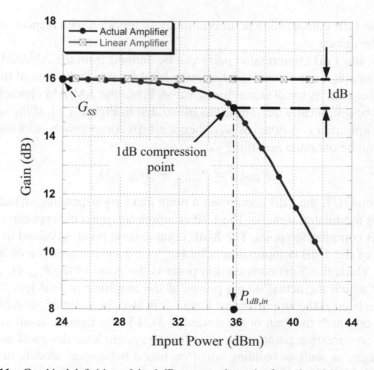

Figure 1.11   Graphical definition of the 1 dB compression point from the AM/AM characteristic

1.6 Third and Fifth Order Intercept Points

The 1 dB compression point characterizes the nonlinear behavior of PAs by only considering the power at the fundamental frequency. However, as amplifiers are driven deeper into their nonlinear regions, the amount of power generated at harmonic and inter-modulation frequencies becomes more significant. The intercept points are defined for odd order harmonics under a single-tone drive signal and odd order inter-modulation products under a multi-tone drive signal as these odd order harmonics and inter-modulation products fall within the close vicinity of the fundamental signal frequency. Third order intercept points are commonly used while fifth order intercepts points are used to a lesser extent. Higher order intercept points are seldom used since the power level generated at their corresponding frequencies is usually too low to have any significant impact on the behavior of the PA.

Figure 1.12 reports, for a sample amplifier driven by a two-tone test signal at frequencies f_1 and f_2 (with $f_1 < f_2$) and having equal amplitudes, the output power at the fundamental frequency (P_{out,f_1}) as a function of the total input power (P_{in}). In this same figure, the output power of the lower third order inter-modulation product ($P_{out,2f_1-f_2}$) is also plotted as a function of the total input power. The linear portion of the P_{out,f_1} vs. P_{in} characteristic has a $1:1$ slope. However, the linear portion of the $P_{out,2f_1-f_2}$ vs. P_{in} characteristic has a $3:1$ slope as it can be deduced from Equation 1.4.

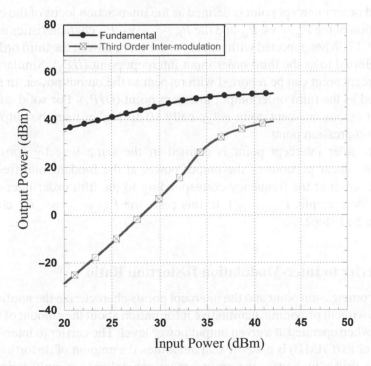

Figure 1.12 Output power characteristics at the fundamental and third order inter-modulation frequencies

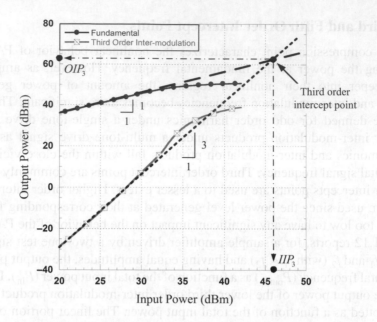

Figure 1.13 Graphical definition of the third order intercept point

The third order intercept point is defined as the intersection locus of the extrapolated linear portion of the P_{out,f_1} vs. P_{in} and the $P_{\mathrm{out},2f_1-f_2}$ vs. P_{in} characteristics as illustrated in Figure 1.13. When reported with respect to the input power, the third order intercept point is referred to as the third order input intercept point (IIP_3). Similarly, the third order intercept point can be reported with respect to the output power. In such a case, it is labeled as the third order output intercept point (OIP_3). For solid state PAs, the third order output intercept point is typically 10 dB higher than the output power at the 1 dB compression point.

The fifth order intercept point is defined in the same way by considering the extrapolated linear portions of the output power at the fundamental frequency and the output power at the frequency corresponding to the fifth order inter-modulation products (for example, $P_{\mathrm{out},3f_1-2f_2}$). In this case, the $P_{\mathrm{out},3f_1-2f_2}$ vs. P_{in} characteristic will have a 5 : 1 slope.

1.7 Carrier to Inter-Modulation Distortion Ratio

The 1 dB compression point and the intercept points characterize the nonlinear behavior of a PA without providing quantitative information about the amount of distortion it generates when operated at a given output power level. The carrier to inter-modulation distortion ratio (C/IMD) is a metric that quantifies the amount of distortion at the output of a PA driven by a two-tone, or in a more general case, a multi-tone test signal.

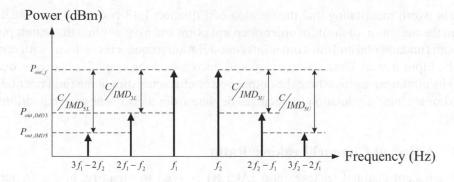

Figure 1.14   Graphical definition of the carrier to inter-modulation distortion ratio

It represents the ratio (in a linear scale) or equivalently the difference (in a logarithmic scale) between the power at the fundamental frequency (carrier) and the power generated at an inter-modulation frequency. The C/IMD is expressed in decibels relative to the carrier (dBc).

For an amplifier driven by a two-tone test signal at frequencies f_1 and f_2 (with $f_1 < f_2$), inter-modulation frequencies of interest commonly are the lower and upper third order inter-modulations ($2f_1 - f_2$ and $2f_2 - f_1$, respectively) and fifth order inter-modulations ($3f_1 - 2f_2$ and $3f_2 - 2f_1$, respectively). Figure 1.14 presents the power spectrum (in dBm) at the output of a memoryless PA having a fifth order nonlinearity and driven by a two-tone test signal at frequencies f_1 and f_2 (with $f_1 < f_2$). This figure graphically defines the lower and upper carrier to third order inter-modulation distortion ratios ($\dfrac{C}{IMD_{3L}}$ and $\dfrac{C}{IMD_{3U}}$, respectively) and those of the fifth order.

In memoryless PAs, the lower and upper C/IMD ratios are equal as reported in Figure 1.14. However, the stronger the memory effects of the amplifier are, the more significant the C/IMD asymmetry will be. The study of the asymmetry between the upper and lower C/IMDs provides an indication of the memory effects exhibited by the PA.

For a memoryless PA, it is possible to predict the third order C/IMD based on the operating output power and the third order intercept point of the device. Using the illustration of Figure 1.13, one can graphically determine that:

$$\frac{C}{IMD_3} = P_{\text{out},f} - P_{\text{out},IMD3} = 2 \times (OIP_3 - P_{\text{out},f}) \tag{1.13}$$

where $\dfrac{C}{IMD_3}$ is the carrier to third order inter-modulation distortion ratio and $P_{\text{out},f}$ and $P_{\text{out},IMD3}$ are the output power levels at the fundamental and third order inter-modulation frequencies, respectively. OIP_3 is the output power at the third order intercept point.

It is worth mentioning that the relation of Equation 1.13 is derived geometrically from the definition of the third order intercept point and assumes that the output power at both fundamental and third order inter-modulation frequencies is linear with respect to the input power. Thus, its accuracy will decrease as the amplifier is driven deeper into its nonlinear region where the output power characteristics at the fundamental and third order inter-modulation frequencies deviate from their linear approximations.

1.8   Adjacent Channel Leakage Ratio

The adjacent channel leakage ratio (ACLR) is used to quantify, in the frequency domain, the nonlinearity of PAs driven by modulated signals. It corresponds to the filtered ratio of the mean power in the main channel to the filtered mean power in an adjacent channel. This is a critical linearity parameter since the power generated by the nonlinear distortions in the adjacent channels cannot be eliminated by filtering and is perceived as interference when the adjacent channels are used for transmission. Thus, the power generated in the adjacent channels is considered as an unwanted emission that needs to be minimized and controlled. Accordingly, each communication standard stipulates, as part of the technical specifications of the transmitter characteristics, the ACLR threshold (also known as the spectrum emission mask) for base stations. A general illustration of the ACLR is illustrated in Figure 1.15, which

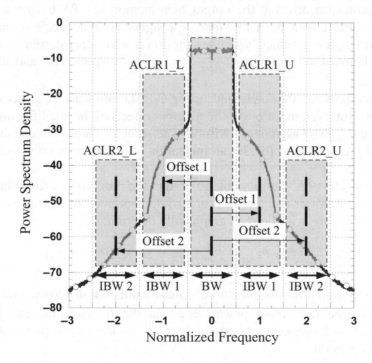

Figure 1.15   Graphical definition of the adjacent channel leakage ratio

reports a sample spectra at the output of the nonlinear transmitter as a function of the normalized frequency. The normalized frequency (f_n) is defined according to:

$$f_n = \frac{f - f_0}{BW} \tag{1.14}$$

where f and f_0 are the absolute frequency and the carrier frequency, respectively. BW represents the bandwidth of the signal.

Figure 1.15 shows that the channel power is calculated in a span that commonly equals the signal bandwidth (BW) and is centered around a normalized frequency of 0 (or equivalently an absolute frequency of f_0). It also shows the ACLR in the lower and upper first adjacent channels (ACLR1_L and ACLR1_U, respectively), and the ACLR in the lower and upper second adjacent channels (ACLR2_L and ACLR2_U, respectively). For each channel, the ACLR calculation requires the definition of the offset frequency that corresponds to the difference between the center of the main channel and that of the considered adjacent channel, as well as the integration bandwidth over which the power will be calculated in the considered adjacent channel.

The parameters used to calculate the ACLR are defined by the communication standards. These parameters include the main channel bandwidth, the adjacent channel parameters (offset frequency and integration bandwidth), and the type and parameters of the filter to be used to calculate the mean power.

1.9   Error Vector Magnitude

The error vector magnitude (EVM) is another measure used to quantify the nonlinear distortions of RF PAs and transmitters. The EVM is defined in the constellation domain and evaluates the deviation between the reference constellation point that should have been obtained in absence of distortions and the actual constellation point obtained in presence of distortions.

Transmitter distortions can be of three types: phase distortions, amplitude distortions, and in the more general cases, simultaneous phase and amplitude distortions. These three cases are illustrated in Figure 1.16 for the constellation diagram of a QPSK (Quadrature Phase Shift Keying) modulation scheme. Phase distortion appears as a rotation of the constellation points causing a phase error as shown in Figure 1.16a. Conversely, amplitude distortions will cause a magnitude error between the amplitudes of the vectors associated with the actual and reference constellation points as depicted in Figure 1.16b. Amplitude and phase distortions will result in an error on both the amplitude and phase of the vector associated with the demodulated constellation point. The effects of simultaneous phase and amplitude distortions on the constellation is illustrated in Figure 1.16c.

In the constellation domain, the error vector refers to the difference between the actual vector of the demodulated constellation point (S_n) and the reference vector associated with the corresponding reference constellation point ($S_{r,n}$) as shown in

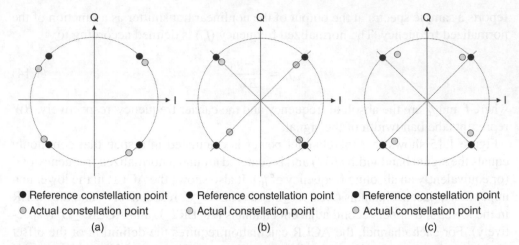

Reference constellation point   ● Reference constellation point   ● Reference constellation point
○ Actual constellation point   ○ Actual constellation point   ○ Actual constellation point

(a)   (b)   (c)

Figure 1.16 Effects of phase and amplitude distortions on the QPSK constellation. (a) Effects of phase distortions. (b) Effects of amplitude distortions. (c) Effects of phase and amplitude distortions

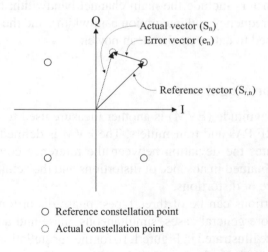

Figure 1.17 Graphical definition of the error vector

Figure 1.17. The EVM refers to the magnitude of the error vector, which is different from the error in the magnitudes except for the particular case where no phase distortions occur.

Threshold EVM values are specified for each communication standard and the latest technical specifications on the transmit modulation quality should be consulted. The EVM is typically expressed in percentage and calculated as the square root of the ratio of the mean power of the error vector to the mean reference power according to:

$$\text{EVM (\%)} = \sqrt{\frac{\frac{1}{N}\sum_{i=1}^{N}|e_i|^2}{\frac{1}{N}\sum_{i=1}^{N}|S_{r,i}|^2}} = \sqrt{\frac{\frac{1}{N}\sum_{i=1}^{N}|S_i - S_{r,i}|^2}{\frac{1}{N}\sum_{i=1}^{N}|S_{r,i}|^2}} \tag{1.15}$$

where N is the number of samples in the waveform. S_i and $S_{r,i}$ are vectors associated with the ith demodulated and reference constellation points, respectively. While e_i is the ith error vector between the demodulated and actual constellation points as defined in Figure 1.17.

References

[1] Lavrador, P., Cunha, T.R., Cabral, P. and Pedro, J.C. (2010) The linearity-efficiency compromise. *IEEE Microwave Magazine*, **11** (5), 44–58.
[2] Katz, A. (2001) Linearization: reducing distortion in power amplifiers. *IEEE Microwave Magazine*, **2** (4), 37–49.
[3] Boulejfen, N., Harguem, A. and Ghannouchi, F.M. (2004) New closed-form expressions for the prediction of multitone intermodulation distortion in fifth-order nonlinear RF circuits/systems. *IEEE Transactions on Microwave Theory and Techniques*, **52** (1), 121–132.
[4] Wu, Q., Xiao, H. and Li, F. (1998) Linear RF power amplifier design for CDMA signals: a spectrum analysis approach. *Microwave Journal*, **41** (12), 22–40.
[5] Pedro, J.C. and Carvalho, N.B. (1999) On the use of multitone techniques for assessing RF components' intermodulation distortion. *IEEE Transactions on Microwave Theory and Techniques*, **47** (12), 2393–2402.
[6] Carvalho, N.B., Remley, K.A., Schreurs, D. and Card, K.C. (2008) Multisine signals for wireless system test and design. *IEEE Microwave Magazine*, **9** (3), 122–138.
[7] Han, S.H. and Lee, J.H. (2005) An overview of peak-to-average power ratio reduction techniques for multi-carrier transmission. *IEEE Wireless Communications*, **12** (2), 56–65.
[8] Jiang, T. and Wu, Y. (2008) An overview: peak-to-average power ratio reduction techniques for OFDM signals. *IEEE Transactions on Broadcasting*, **54** (2), 257–268.
[9] Rahmatallah, Y. and Mohan, S. (2013) Peak-to-average power ratio reduction in OFDM systems: a survey and taxonomy. *IEEE Communications Surveys & Tutorials*, **15** (4), 1567–1592.
[10] Hammi, O. and Ghannouchi, F.M. (2009) Power alignment of digital predistorters for power amplifiers linearity optimization. *IEEE Transactions on Broadcasting*, **55** (1), 109–114.

2

Dynamic Nonlinear Systems

In Chapter 1 a description of a nonlinear system, the power amplifier, along with the effects that it introduces to the communication signal, is presented. Characteristics of this nonlinear system were also presented in detail along with the metrics to quantify the amount of nonlinear distortion. In this chapter, dynamic nonlinear power amplifiers will be introduced. First, the notion of memory in systems will be defined. Then a classification of nonlinear power amplifier systems based on their amount of memory will be provided. The origins of the linear and nonlinear memory effects and their characteristics will be addressed. A general model based on the Volterra series to model power amplifiers with memory effects will be introduced. Its pass-band time domain representation and its baseband equivalent model will be also provided.

2.1 Classification of Nonlinear Systems

Memory in systems can be defined as the ability of a system to behave as a function of values of the input signal that are different than the present value of the input signal. Therefore, systems can be classified in two groups: memoryless systems and systems with memory [1–4].

2.1.1 Memoryless Systems

In general, a system is said to be memoryless if its output at a given time t_o, $y(t_o)$, is a function of only the input value at t_o, $x(t_o)$.

For practical considerations, every physical system is causal, which means that $y(t_o)$ cannot be a function of future values of $x(t_o)$. Moreover, every physical system will certainly introduce a certain delay to the input signal. Since this delay will not affect the integrity of the signal, it can be assumed without loss of generality that a system is memoryless if its output at a given time t_o, $y(t_o)$, is a function of only one input value

Behavioral Modeling and Predistortion of Wideband Wireless Transmitters, First Edition.
Fadhel M. Ghannouchi, Oualid Hammi and Mohamed Helaoui.
© 2015 John Wiley & Sons, Ltd. Published 2015 by John Wiley & Sons, Ltd.

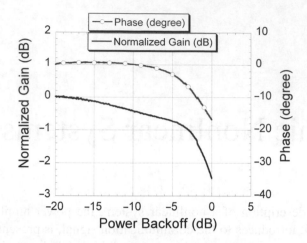

Figure 2.1 Example of an AM/AM and AM/PM curve for a memoryless system

$x(t_o - \tau)$, where τ is the delay introduced by the system.

$$y(t_o) = f[x(t_o - \tau)] \tag{2.1}$$

From Equation 2.1, it can be concluded that for similar values of input signal, the corresponding outputs are similar as well. Therefore, the curve of the output signal versus the input signal for a memoryless system is a single line. An example of such a curve is shown in Figure 2.1.

2.1.2 Systems with Memory

In general, a system is said to have memory if its output at a given time t_o, $y(t_o)$, is a function of inputs other than $x(t_o)$.

For practical considerations, given that every physical system is causal, this will certainly introduce a certain delay τ to the input signal, it can be assumed without loss of generality that a physical system has memory if its output at a given time t_o, $y(t_o)$, is a function of values of input signal preceding $x(t_o - \tau)$.

$$y(t_o) = f[x(t_o - \tau), x(t_o - \tau - \tau')] \tag{2.2}$$

where τ' may be any constant to show that the output can be also a function of any other past samples of the input signal.

From Equation 2.2, it can be concluded that for similar values of input signal, the corresponding outputs may be different. Therefore, the curve of the output signal versus the input signal for a system with memory is not a single line. An example of such a curve is shown in Figure 2.2.

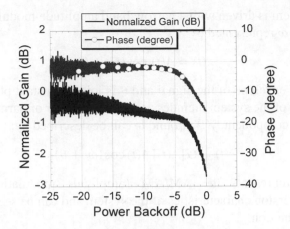

Figure 2.2   Example of an AM/AM and AM/PM curve for a system with memory

2.2   Memory in Microwave Power Amplification Systems

All communication systems, in particular transmitters, have inherent nonlinearities that limit their usefulness and range of applications. For example, the input power level in microwave amplifiers must be kept below a certain level to ensure operation in a region of sufficiently linear amplification. Ignoring this requirement leads to the generation of significant intermodulation products caused by amplitude and phase nonlinearities. The types of nonlinear systems can be briefly classified as:

- Nonlinear systems without memory
- Nonlinear systems effectively without memory
- Nonlinear systems with memory.

Each type of system produces distinct nonlinear effects. These three types of systems and their effects are characterized in the following subsections [5–8].

2.2.1   Nonlinear Systems without Memory

Systems belonging to this category have the following three characteristics:

- The output instantaneously responds to the input
- The system does not have a frequency response
- There are no phase nonlinearities.

Nonlinearities without memory are sometimes called resistive nonlinearities. Indeed, a nonlinear circuit without energy storage elements cannot possess memory.

When such a system is driven with a narrow band amplitude modulated signal $\tilde{x}(t)$ at carrier frequency ω represented by:

$$\tilde{x}(t) = A(t) \cos(\omega t + \theta) \tag{2.3}$$

where, $A(t)$ is the envelope of the signal and θ is initial constant phase of the signal. The output signal of the system includes an infinite number of harmonic components and the bandpass component, $\tilde{y}(t)$, around ω can be described by:

$$\tilde{y}(t) = G[A(t)]A(t) \cos(\omega t + \theta) \tag{2.4}$$

where $G[A(t)]$ represents the AM/AM (amplitude modulation to amplitude modulation) conversion characteristics of the system and can be seen as an envelope-dependent gain function.

A necessary requirement for inclusion in the memoryless category is that $G[A(t)]$ should not depend on frequency. In other words, the magnitude response of the system is "flat" in the frequency domain. The effects of memoryless nonlinearities are:

- Generation of nonlinear amplitude distortion,
- Generation of harmonic frequencies and intermodulation products,
- A possible shift in the system's DC operating point due to even-order distortions.

Examples of this type of nonlinearity are the piecewise-linear limiter and the ideal comparator. An appropriate representation for such characteristics is the relatively simple, classic power (Taylor) series; for this reason, series based formulations are often used in nonlinear modeling of systems. It should be understood that no real system can ever be truly without memory due to the always-present reactive (capacitive and inductive) elements in any electronic circuits.

2.2.2 Weakly Nonlinear and Quasi-Memoryless Systems

Systems in this category exhibit nonlinear amplitude modulation to phase modulation (AM/PM) conversion behavior and AM/AM conversion behavior. Furthermore, both the nonlinear AM/AM and AM/PM characteristics of the system do not have a measurable frequency dependency, due to either (or both) of the following causes:

- The input signals are limited to narrow band modulated signals around a carrier frequency, ω, around which no significant frequency response is observed for the AM/AM and AM/PM behaviors of the system,
- The nonlinear transfer functions simply do not depend on frequency.

An important implication of this system requirement is that phase nonlinearities can be present, but no frequency response is allowed. Such a system therefore represents a cross domain between the memoryless system and a full-memory nonlinear system. When such a system is driven with narrow band amplitude modulated signal, $\tilde{x}(t)$, at carrier frequency ω represented by:

$$\tilde{x}(t) = A(t) \cos[\omega t + \theta(t)] \tag{2.5}$$

The output signal of the system includes an infinite number of harmonic and inter-modulation components and the bandpass component, $\tilde{y}(t)$, around ω can be described by:

$$\tilde{y}(t) = G[A(t)]A(t) \cos\{\omega t + \theta(t) + \phi_G[A(t)]\} \tag{2.6}$$

where $G[A(t)]$ represents the AM/AM conversion characteristic and $\phi_G[A(t)]$ is the AM/PM conversion characteristic and both can be seen as an envelope-dependent complex gain function.

2.2.3   Nonlinear System with Memory

This last category is the most general, as it includes the previous two categories as special cases. All of the nonlinear effects of the previous two categories are still present, but the additional property of frequency dependence in the AM/AM and AM/PM coefficients may be observed.

When such a system is driven with narrow band amplitude modulated signal, $\tilde{x}(t)$, at carrier frequency ω represented by:

$$\tilde{x}(t) = A(t) \cos[\omega t + \theta(t)] \tag{2.7}$$

The output signal of the system includes an infinite number of harmonic and inter-modulation components and the bandpass component, $\tilde{y}(t)$, around ω can be described by:

$$\tilde{y}(t) = G[A(t), \omega]A(t) \cos\{\omega t + \theta(t) + \phi_G[A(t), \omega]\} \tag{2.8}$$

Proper modeling requires that attention be paid to the frequency characteristics of the nonlinearities. The Volterra series is an appropriate representation, although frequency-dependent in-phase/quadrature models have been proposed.

Potential applications for nonlinear models incorporating memory include broad-band amplifiers, where the input signal is spread over a wide frequency range. This would include all TWTAs (Traveling Wave Tube Amplifiers) and SSPAs (solid-state power amplifiers).

2.3   Baseband and Low-Pass Equivalent Signals

In practice, for modeling or analysis purposes of relatively low frequency modulated signals, baseband signals are considered. Baseband signals have frequency spectra concentrated near zero frequency. However, for wireless communications where, in theory, the carrier frequency, f_c, is relatively high in the gigahertz range, most of the time pass-band signals are considered and used for the purpose of simulation of wireless systems. Pass-band signals have frequency spectra concentrated around the carrier frequency. Baseband signals can be converted to pass-band signals through down-conversion and *vice versa* through up-conversion [7, 8].

In wireless communication systems, a baseband signal is up-converted to a bandpass signal by amplitude, phase, or frequency modulation, so that it can be transmitted. The amplitude and phase modulated bandpass signal can be described as:

$$\tilde{x}(t) = A(t)\cos[\omega_c t + \theta(t)] \tag{2.9}$$

where $\omega_c = 2\pi f_c$ is the angular carrier frequency and $A(t)$ and $\theta(t)$ are the amplitude and phase signals that modulated the carrier, respectively.

The signal described in Equation 2.9 has an envelope bandwidth much lower than the carrier frequency and is called a *bandpass signal with center frequency, f_c*. Using trigonometric identities, this signal can be written as:

$$\tilde{x}(t) = A(t)\cos[\theta(t)]\cos(\omega_c t) - A(t)\sin[\theta(t)]\sin(\omega_c t)$$

$$= I(t)\cos(\omega_c t) - Q(t)\sin(\omega_c t) \tag{2.10}$$

where $I(t)$ denotes the in-phase component and $Q(t)$ is the quadrature component, which are defined as:

$$I(t) = A(t)\cos[\theta(t)] \tag{2.11}$$

$$Q(t) = A(t)sin[\theta(t)] \tag{2.12}$$

Equation 2.9 can be written in complex form as:

$$\tilde{x}(t) = \text{Re}[A(t)e^{j\omega_c t}e^{i\theta(t)}] = \text{Re}[x(t)e^{j\omega_c t}] \tag{2.13}$$

where $x(t)$ is called the baseband signal or complex envelope and contains the same information as the bandpass signal $\tilde{x}(t)$ and can be represented as

$$x(t) = I(t) + jQ(t) = A(t)e^{i\theta(t)} \tag{2.14}$$

On the receiver side, the baseband signal $x(t)$ can be obtained from the bandpass signal $\tilde{y}(t)$ through down conversion, demodulation, and following the channel equalization process as shown in Figure 2.3. $\tilde{y}(t)$ represents the signal at the output of the power amplifier/transmitter.

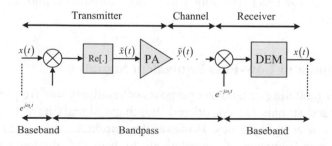

Figure 2.3   Pass-band and baseband signals

2.4 Origins and Types of Memory Effects in Power Amplification Systems

2.4.1 Origins of Memory Effects

Memory effects can be explained simply by the fact that the output of the system exhibiting memory at any instant is a function not only of the corresponding instantaneous input (after compensating for the system delay) but also of the inputs at other instants (past instants for causal systems). This is an inherent characteristic of energy-storing circuits or elements of the memory system. In the case of power amplifiers, intrinsic, and extrinsic parasitic elements, matching network elements, and the nature of the transistor junction might include energy-storing circuits or elements that will result in memory effects.

Depending on the correlation with the nonlinearity of the transistor, one can classify these memory effects into two categories [7, 11–15].

1. *Linear memory effects*, which are memory behaviors uncorrelated with the nonlinear response of the power amplifier. They are generally represented mathematically as a linear combination of the input signal at different time shifts. For instance, if a system exhibits only linear memory effects, its output can be expressed as:

$$y(t) = \sum_i h_i x(t - \tau_i) \tag{2.15}$$

   This is the expression of finite impulse response filters, which are linear systems.

   While power amplifiers are nonlinear systems, their output include two terms, a linear term and a nonlinear term. The linear term represents a linear behavior including linear memory effects.

2. *Nonlinear memory effects*, which are memory behaviors mixed with the nonlinearity of the transistor. While the source of these memory effects may be linear circuits, such as capacitors, for example, the combination of the memory effect of these linear circuits with the nonlinear behavior of the transistor results in a term in the output signal of the power amplifier that includes a nonlinear function of different samples of the input signal at different instances.

   The output of a nonlinear power amplifier can be expressed as:

$$y(t) = \underbrace{\sum_i h_i x(t - \tau_i)}_{\text{linear term including linear memory effects}} + \underbrace{f_{nonlinear}[x(t - \tau_1), \ldots, x(t - \tau_N)]}_{\substack{\text{nonlinear term including nonlinearity} \\ \text{and nonlinear memory effects}}} \tag{2.16}$$

Memory effects can also be classified in two categories based on their origins [7, 16–18]:

1. One can distinguish memory effects caused by the active device's temperature modulation. This category of memory effects is called electro-thermal or thermal

memory effects. Given it is function of the temperature change in the junction of the transistor, this category of memory effect has a long term effect and affects narrow bandwidths of the signal spectrum.

2. The second category consists of electrical memory effects, which are produced by the external terminations, including parasitic elements and matching networks of the power amplifier. The properties of these terminations across the fundamental frequency, baseband frequency, and all the harmonic frequencies shape the power amplifier response around the carrier frequency.

In order to be able to analyze the memory effects and model them properly, it is important to understand the different circuits in a power amplifier, their behaviors, and their characteristics. In the following, for each origin of memory effects, these circuits are modeled and explained. Their effect on the power amplifier behavior is then discussed and compared with a memoryless behavior.

2.4.2 Electrical Memory Effects

The main origins of electrical memory effects are the transistor terminations, including intrinsic and extrinsic parasitic elements, and matching networks. In order to better analyze the electrical memory effects, it is important to understand the impedance termination in transistor amplifiers [19, 20].

Figure 2.4 shows a block diagram of a common source MESFET (Metal Semiconductor Field Effect Transistor) amplifier. Z_{G_match} is the impedance presented by the input matching network, excluding the biasing network, to the source of the gate of the transistor, Z_{G_bias} is the impedance presented by the biasing network to the gate of the transistor, and Z_{G_in} is the impedance presented by looking at the gate of the transistor. Similarly, at the output of the transistor, the impedance presented by the drain

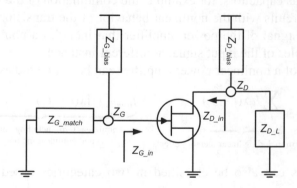

Figure 2.4 Block diagram of a common source MESFET amplifier showing the definition of the different impedances

is Z_{D_in}, the impedance presented by the biasing network is Z_{D_bias}, and the impedance presented by the loading or output matching network to the drain of the transistor is Z_{Dn_L}. The impedances of gate and drain nodes can then be obtained by:

$$Z_G = Z_{G\_match}//Z_{G\_bias}//Z_{G\_in} \tag{2.17}$$

$$Z_D = Z_{D\_in}//Z_{D\_bias}//Z_{D\_L} \tag{2.18}$$

Given that the transistor impedances Z_{G_in} and Z_{D_in} vary as a function of the driving signal power level and operating conditions of the transistor, the transistor will exhibit a nonlinear behavior at the gate and drain levels. It can then be concluded that nonlinear power amplifiers may include more than one nonlinear element. The simplest model of a nonlinear power amplifier will include:

1. A nonlinear block representing the gate voltage as a function of the input signal to the power amplifier.
2. A nonlinear block representing the relationship between the gate voltage and drain voltage.

Each of these two blocks also includes a frequency response that is due to the transistor behavior variation versus frequency and matching network response versus frequency at the fundamental frequency and each of the harmonic frequencies. The cascade of these nonlinear elements results in mixing the linear memory effects (frequency responses around a carrier frequency or its harmonic) along with the nonlinear behaviors of the nonlinear elements. This mixing of linear memory effects and nonlinear response will result in an output signal that includes nonlinearity along with nonlinear memory effects around the fundamental carrier of the signal. These nonlinear memory effects include products that are function of the frequency response of the matching network and transistor not only at the fundamental frequency but also around the different harmonic, which are translated to the fundamental frequency via the nonlinear elements [14, 21, 22].

In order to understand this concept, one can simplify the modeling of the transistor to a cascade of two nonlinear systems, G and H, each having linear memory in the form of a frequency response at each of the fundamental and harmonic frequencies. Figure 2.5 shows a block diagram of this cascade and illustrates the origins of the intermodulation products at the output and how they are affected by the frequency response of the system at the fundamental and harmonic frequencies. Each of the two systems is modeled by a nonlinearity in the order of three and a set of frequency responses around each of the fundamental and carrier frequencies (G_0, G_1, G_2, and G_3 are the frequency responses of G around the envelope, the fundamental, second and third harmonics, respectively; and H_0, H_1, H_2, and H_3 are the frequency responses of H around the envelope, the fundamental, second, and third harmonics, respectively). The

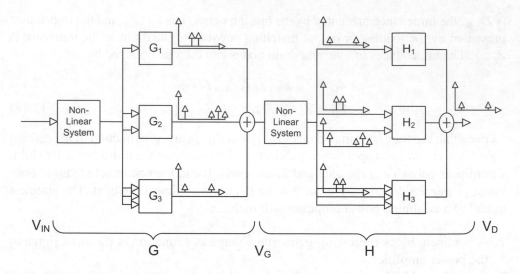

Figure 2.5   Modeling of nonlinear electrical memory effects in a cascade of two nonlinear systems

third order intermodulation products at the output of the system are the combination of different products including products generated by:

- The third order nonlinearity of the first system, G, passed through the frequency response H_1 around the fundamental carrier frequency, of the second system, H.
- The second order mixing product of the fundamental output, and the envelope and the second harmonic outputs of the first system, G, which also passes through the frequency response, H_2, around the second harmonic in the second system, H.
- The third order mixing product of the fundamental output of the first block, G, which also passes through the frequency response, H_3, around the second harmonic in the second system, H.

This third intermodulation product at the output of the power amplifier is a function of different nonlinearity orders including even nonlinearity orders and frequency responses at the envelope frequency, the fundamental frequency, and different harmonic frequencies.

If the signal bandwidth is W, and by only considering nonlinearities up to the third order, the nonlinear memory effect is a transistor is affected by:

- The frequency response along a band of W around DC frequency,
- The frequency response along a band of W around the fundamental frequency,
- The frequency response along a band of $2W$ around the second harmonic frequency, and
- The frequency response along a band of $3W$ around the third harmonic frequency.

For practical considerations, on one hand, the frequency responses around the fundamental, second harmonic, and third harmonic frequencies are considered to occur around the same fractional bandwidth and are generally insignificant for single carrier and relatively narrowband applications. Their effect may be of importance if multi-carrier and significantly wideband signals are considered. On the other hand, the frequency response around DC frequency will have significant effect on the memory even for relatively narrowband applications if no careful design of the biasing circuit is carried out in order to maintain constant gate node impedance in this frequency band.

2.4.3   Thermal Memory Effects

As it is indicated by its name, the thermal memory effect is caused by the electrothermal coupling in the power transistor. It is a function of the power dissipated in the transistor, which directly affects the temperature of the transistor junction. As a result, the characteristics of the transistor in terms of gain and output power capability change versus these temperature variations. Given the fact that the temperature will vary more slowly than the amplitude of the signal variation, the thermal memory effect manifests and usually impacts the low frequency components of the signal below the MHz range. To analyze the thermal memory effect in transistors, one should first analyze the power dissipation and temperature change in the power amplifier circuit [23–26].

The power dissipation in a FET (Field Effect Transistor) operated in normal conditions (gate current equals to zero) is provided by:

$$P_{dissipated}(t) = v_{ds}(t) \cdot i_{ds}(t) \tag{2.19}$$

where $v_{ds}(t)$ is the drain-source voltage and $i_{ds}(t)$ is the drain current of the transistor.

In order to analyze the temperature variation in the transistor junction, thermal impedance, Z_{th}, is defined as the ratio between the temperature rise and heat flow from the device. Figure 2.6a shows the heat dissipation in a power transistor, from the device chip to the heat sink passing through the package of the device and circuit board. Given that the heat dissipation from one stage to another is not instantaneous, a delay and discharging behavior can be modeled. These effects will result in a non-purely resistive thermal impedance model. In this model, thermal resistances, R_{th}, describe a steady-state behavior of the temperature while thermal capacitances, C_{th}, describe the dynamic behavior. Together, thermal resistors, R_{th}, and thermal capacitors, C_{th}, result in modeling temperature variation with a giving rising and falling constant $R_{th}C_{th}$.

Using complex thermal impedances for each connection, this heat dissipation can be modeled by a set of lumped elements forming a low-pass filter topology as shown in Figure 2.6b. This modeling is in agreement with the expectations that we presented earlier in this section, which consists of the fact that thermal memory effect affects low frequency components of the signal. In practice, the low-pass filter topology will have a bandwidth varying between 100 kHz and 1 MHz depending on the nature of

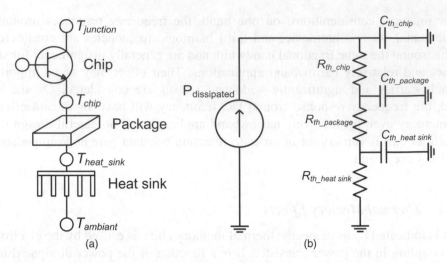

(a)   (b)

Figure 2.6 Modeling of thermal memory effects in a power transistor. (a) Different temperatures definitions. (b) Circuit modeling of the temperature variation

the chip connection to the heat sink. The thermal memory effect then affects signal
frequency components lower than 1 MHz.

Using the circuit modeling of Figure 2.6a, the temperature variation in the transistor
junction can then be given by:

$$\Delta T = T_{junction} - T_{ambiant} = P_{dissipated} \cdot Z_{th} \qquad (2.20)$$

More precisely, the thermal impedance may vary versus frequency and the dissipated power has different frequency components. In fact, the dissipated power is the
product of two signals around DC and the fundamental frequency. Such product will
have components around DC, the envelope, the fundamental frequency, and second
harmonic. The products around the fundamental frequency and the second harmonic
are filtered out by the filter topology and only the DC and envelope components affect
the temperature variation of the power transistor junction. Equation 2.20 can then be
rewritten as:

$$\Delta T = T_{junction} - T_{ambient}$$
$$= P_{dissipated} (0\,\text{Hz}) \cdot R_{th}(0\,\text{Hz}) + P_{dissipated} (f_1 - f_2) \cdot Z_{th}(f_1 - f_2) \qquad (2.21)$$

where f_1 and f_2 are two different frequencies within the band of the modulated signal, $P_{dissipated}(f_1 - f_2)$ is the part of the envelope component of the dissipated power
corresponding the two different frequencies f_1 and f_2. $R_{th}(0\,Hz)$ and $Z_{th}(f_1 - f_2)$ are
the thermal resistor value at zero frequency and the thermal impedance at frequency
$(f_1 - f_2)$; and $p_{dissipated}(0\,Hz)$ is the DC component of the power dissipation.

The temperature variation expression of Equation 2.21 includes two terms. The first
term is related to the DC dissipation and is frequency independent. The second term

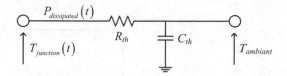

Figure 2.7   Simplified transistor thermal modeling

is function of the envelope dissipation and is frequency dependent, which means that any change in the transistor characteristics due to the temperature variation results in frequency dependent effects or memory effects.

Much research work has investigated the temperature variation in the power transistor junction for different types of signals in order to model it and understand its effect on the generation of thermal memory effects. These activities contributed to proper modeling and linearization of the thermal memory effects. To understand better how the junction temperature of the power transistor varies as a function of the input signal and how this will affect the signal integrity, an analysis and modeling of the transistor thermal behavior in the presence of a pulsed signal is given next [27].

First, given the fact that the thermal constants related to the heat sink dissipation, $R_{th\_heat\ sink}$, and $C_{th\_heat\ sink}$, are too large compared to the package and chip thermal constants, $R_{th\_package}$, $C_{th\_package}$, R_{th_chip}, and C_{th_chip}, the temperature of the heat sink, T'_{th}, is almost equal to the ambient temperature – and hence it can be considered independent from signal variations. One can therefore ignore the effect of $R_{th\_heat\ sink}$ and $C_{th\_heat\ sink}$. Moreover, in order to further simplify the analysis, one can model the joint effect of the heat dissipation in the chip and package by a set of an equivalent thermal resistor R_{th} and an equivalent thermal capacitor C_{th}. This results in simplifying the circuit in Figure 2.6b to the circuit in Figure 2.7. Using this simplified modeling circuit for the junction temperature variation, the relationship between the junction temperature $T_{junction}$ and the ambient temperature $T_{ambient}$ is given by:

$$\frac{\partial T_{junction}(t)}{\partial t} + \frac{1}{R_{th}C_{th}}T_{junction}(t) = \frac{1}{R_{th}C_{th}}[R_{th} \cdot P_{dissipated}(t) + T_{ambient}] \qquad (2.22)$$

where $p_{dissipated}(t)$ defined in Equation 2.19 can also be expressed as a function of the output power, $P_{RF\_out}(t)$, and instantaneous power efficiency, $\eta(t)$, of the power amplifier by:

$$P_{dissipated}(t) = [1 - \eta(t)] \cdot P_{RF\_out}(t) \qquad (2.23)$$

Equation 2.22 is a first order non-homogenous differential equation that has a general solution in the form of:

$$T_{junction}(t) = C_1 e^{-\frac{t}{\tau}} + \frac{1}{\tau}e^{-\frac{t}{\tau}}\int_1^t e^{\frac{\xi}{\tau}}[R_{th} \cdot P_{dissipated}(\xi) + T_{ambient}]\partial\xi \qquad (2.24)$$

where $\tau = R_{th}C_{th}$ and C_1 is a constant that can be determined by initial conditions.

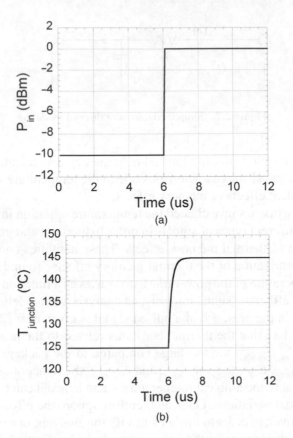

Figure 2.8 Junction temperature variation for a step input. (a) Input signal variation versus time. (b) Corresponding junction temperature variation versus time

If the driving signal in the power amplifier is a step input signal, as shown in Figure 2.8a, the dissipated power follows a step signal shape as well (see Figure 2.8b) and can be given by:

$$P_{dissipated}(t) = \begin{cases} P_H & t > t_o \\ P_L & t < t_o \end{cases} \qquad (2.25)$$

In this case, it can be easily shown that, if $\tau \ll t_o$, the junction temperature expression in Equation 2.24 becomes:

$$T_{junction}(t) = T_{junction,H} + (T_{junction,L} - T_{junction,H})e^{-\frac{t-t_o}{\tau}} ; \; t > t_o \qquad (2.26)$$

where $T_{junction,L} = R_{th}P_L + T_{ambient}$ and $T_{junction,H} = R_{th}P_H + T_{ambient}$.

Using similar reasoning, and by noting that the mathematical formulation will be the same independently from the sign of $(P_H - P_L)$, it can be concluded that if the driving input of transistor is a pulsed signal with period $T_o \gg \tau$ as shown in Figure 2.9a, the junction temperature will have the form of Figure 2.9b.

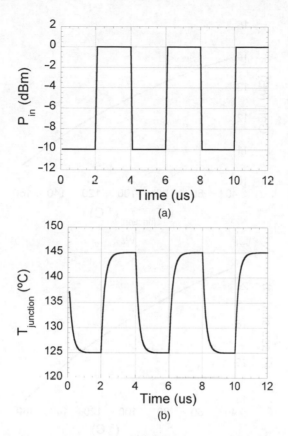

Figure 2.9 Junction temperature variation for a pulsed input. (a) Input signal variation versus time. (b) Corresponding junction temperature variation versus time

The junction temperature variation can be obtained from Equation 2.26 as follows:

$$\Delta T_{junction} = T_{junction,H} - T_{junction,L} = R_{th} \cdot \Delta P_{disspated} \tag{2.27}$$

where $\Delta P_{disspated} = P_H - P_L$ is the maximum variation in the instantaneous power dissipation in the transistor.

The variation in the junction temperature as a function of the signal level results in changes in the power amplifier complex gain, which result in distortion related to thermal memory effect. Indeed, Figure 2.10 shows the variation of the measured complex gain versus the junction temperature for a power amplifier using a 90-W LDMOS (Laterally Diffused Metal Oxide Semiconductor) transistor.

Higher temperatures result in lower gain. Therefore, in the case of a pulsed signal, when transiting from a low to a high level, the junction temperature is low and the gain is higher. During the high level cycle, the junction temperature rises exponentially and the gain drops accordingly. Similarly, when transiting from a high level to a low level, the junction temperature is high and the gain is low. During the low level

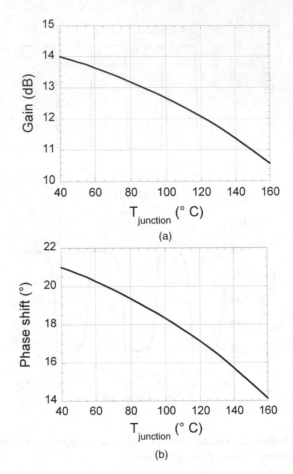

Figure 2.10   Complex gain variation as a function of the junction temperature. (a) Gain in dB. (b) Phase shift in degrees

cycle, the junction temperature drops exponentially and gain increases accordingly. This behavior is shown in Figure 2.11, which shows how the output power is distorted compared to the input power for a power amplifier with pulsed input. This distortion is caused uniquely by thermal memory effects.

2.5   Volterra Series Models

After understanding the origins of memory effects, it is important to take them into consideration when analyzing the effect of power amplifiers on signal linearity. These effects are dependents on different factors related to the signal and power amplifier characteristics. While electrical memory effects are a function of the signal bandwidth, thermal memory effects are a function of the amount of power dissipation in the power amplifier and the cooling circuit used to dissipate the heat from this dissipated power.

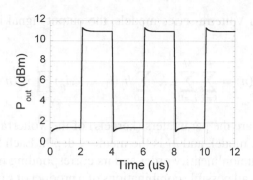

Figure 2.11 Generic block diagram for behavioral model performance assessment

On one side, if a signal with relatively narrow bandwidth is used such that the responses of the amplifier and its matching networks around the envelope frequency are considered constant over the bandwidth of the signal, the output signal of the power amplifier is considered to have non-significant electrical memory effects. The electrical memory effect can then be neglected without affecting the analysis of the power amplifier behavior. In practice, if the signal has a bandwidth lower than 10 MHz, a careful power amplifier design guarantee nearly constant frequency response around the envelope frequency, the fundamental, and the harmonics. In this case, the electrical memory effect can be neglected.

On the other side, if the power amplifier's junction temperature variation is small enough so that its effect on the gain of the power amplifier is insignificant, the thermal memory effect can be neglected. In practice, a temperature variation of few degrees will not introduce significant changes in the power amplifier gain. Therefore, to have negligible thermal memory effect, the variation in the power amplifier power dissipation defined in Equation 2.27 should satisfy: $\Delta P_{disspated}$ should be in the same order of magnitude or smaller than $\frac{1}{R_{th}}$. This condition can be satisfied automatically for ideal class A power amplifiers where $\Delta P_{disspated}$ is zero or for power amplifiers with low power dissipation, for example, efficient switching mode power amplifiers.

If a power amplifier has negligible electrical and thermal memory effects, it can be modeled using a nonlinear static model that does not have any memory effect. Often, Taylor series are used for such modeling and the output of the power amplifier, $x_{out}(n)$, can then be related to the input $x_{in}(n)$ by [3] and [22]:

$$x_{out}(n) = \sum_{i=1}^{K} a_i \cdot x_{in}^i(n) \tag{2.28}$$

However, if the bandwidth increases, the electrical memory effects can no longer be neglected. Therefore, memory effects should be taken into consideration when modeling power amplifiers. The Volterra series [23–25], can be used to accurately characterize a dynamic nonlinear system including linearity and the different types of memory effect. In Volterra series models, the output signal is related to the input signal as follows:

$$x_{out}(n) = \sum_{k=1}^{K} \sum_{i_1=0}^{M} \cdots \sum_{i_p=0}^{M} h_p(i_1, \ldots, i_p) \prod_{j=1}^{k} x_{in}(n - i_j) \tag{2.29}$$

where $h_p(i_1, \ldots, i_p)$ are the parameters (kernels) of the Volterra model, K is the non-linearity order of the model, and M is the memory depth. Each kernel of the Volterra series models a given nonlinearity order and its corresponding memory effect. A k-th order kernel includes all possible combinations of a product of k time shifts of the input signal. Therefore, it includes all possible forms of memory effect and is considered the most complete model to take into account linearity and any type of memory effect. However, it results in a large number of coefficients that increases exponentially with the degrees of the nonlinearity and memory depth of the system. The increase in the number of coefficients increases the computational complexity of the model. Therefore, in practice, the Volterra series model is limited to modeling systems with low nonlinearity and memory orders. To overcome the computational complexity of the Volterra series, different reductions of the Volterra series have been proposed [32, 33]. These complexity reduced models will be described in the following chapters.

References

[1] Haykin, S. and Van Veen, B. (2014) *Signals and Systems*, John Wiley & Sons, Inc, Hoboken, NJ.
[2] Philips, C., Parr, J. and Riskin, E. (2014) *Signals, Systems and Transforms*, 5th edn, Prentice-Hall.
[3] Oppenheim, A. and Willsky, A. (1996) *Signals and Systems*, 2nd edn, Prentice-Hall.
[4] Sundararajan, D. (2008) *A Practical Approach to Signals and Systems*, John Wiley & Sons, Inc, Hoboken, NJ.
[5] Gharaibeh, K. (2012) *Nonlinear Distortion in Wireless Systems*, John Wiley & Sons, Inc, Hoboken, NJ.
[6] Vuolevi, J.H.K., Rahkonen, T. and Manninen, J.P.A. (2001) Measurement technique for characterizing memory effects in RF power amplifiers. *IEEE Transactions on Microwave Theory and Techniques*, **49** (8), 1383–1389.
[7] Vuolevi, J.H.K. and Rahkonen, T. (2002) *Distortion in RF Power Amplifiers*, Artech House.
[8] Vuolevi J.H.K. , Analysis, measurement and cancellation of the bandwidth and amplitude dependence of intermodulation distortion in RF power amplifiers. Doctoral thesis, University of Oulu, Oulu, 2001.
[9] Benedetto, S. and Biglieri, E. (1999) *Principles of Digital Transmission with Wireless Applications*, Plenum Series in Telecommunications, Kluwer Academic Publishers.
[10] Haykin, S. (2013) *Communications Systems*, 5th edn, John Wiley & Sons, Inc, Hoboken, NJ.
[11] Maas, S. (1995) Third-order intermodulation distortion in cascaded stages. *IEEE Microwave and Guided Wave Letters*, **5** (6), 189–191.
[12] T. Rahkonen and J. H. K. Vuolevi, Memory effects in analog predistorting linearizing systems. Proceedings 1999 NORCHIP Conference, Oslo, Norway, November 1999, pp. 114–119, 1999.
[13] Maas, S. (1997) *Nonlinear Microwave Circuits*, IEEE Press, Piscataway, NJ.
[14] Saleh, A.A.M. (1981) Frequency-independent and frequency-dependent nonlinear models of TWT amplifiers. *IEEE Transactions on Communications*, **29** (11), 1715–1720.

[15] R. Raich, and G. T. Zhou, On the modeling of memory nonlinear effects of power amplifiers for communication applications. IEEE Digital Signal Processing Workshop, Pine Mountain, GA, October 2002, pp. 7–10, 2002.

[16] Y. Zhu, J. K. Twynam, M. Yagura, M. Hasegawa, T. Hasegawa, Y. Eguchi, et al., Analytical model for electrical and thermal transients of self-heating semiconductor devices, *IEEE Transactions on Microwave Theory and Techniques*, **46**, 12, 2258–2263, 1998.

[17] Raab, F.H., Asbeck, P., Cripps, S. *et al.* (2002) Power amplifiers and transmitters for RF and microwave. *IEEE Transactions on Microwave Theory and Techniques*, **50** (3), 814–826.

[18] Clark, C.J., Chrisikos, G., Muha, M.S. *et al.* (1998) Time-domain envelope measurement technique with application to wideband power amplifier modeling. *IEEE Transactions on Microwave Theory and Techniques*, **46** (12), 2531–2540.

[19] Gonzalez, G. (1997) *Microwave Transistor Amplifiers: Analysis and Design*, Prentice-Hall, Englewood Cliffs, NJ.

[20] J. H. K. Vuolevi and T. Rahkonen, The effects of source impedance on the linearity of BJT common-emitter amplifiers. Digest 2000 IEEE International Symposium on Circuits and Systems, Geneva, Switzerland, May 2000, pp. 197–200, 2000.

[21] Kim, J. and Konstantinou, K. (2001) Digital predistortion of wideband signals based on power amplifier model with memory. *Electronics Letters*, **37** (23), 1417–1418.

[22] Bosch, W. and Gatti, G. (1989) Measurement and simulation of memory effects in predistortion linearizers. *IEEE Transactions on Microwave Theory and Techniques*, **37** (12), 1885–1890.

[23] E. Schurack, W. Rupp, T. Latzel and A. Gottwald, Analysis and measurement of nonlinear effects in power amplifiers caused by thermal power feedback. Digest IEEE International Symposium on Circuits and Systems, San Diego, CA, May 1992, pp. 758–761, 1992.

[24] T. Hopkins and R. Tiziani, Transient thermal impedance considerations in power semiconductor applications. Proceedings in Automotive Power Electronics, Dearborn, MI, August 1989, pp. 89–97, 1989.

[25] N. Le Gallou, J. M. Nebus, E. Ngoya, and H. Buret, Analysis of low frequency memory and influence on solid state HPA intermodulation characteristics. Digest IEEE International Microwave Symposium, Phoenix, AZ, June 2001, pp. 979–982, 2001.

[26] Fox, R.S. and Zweidinger, D. (1993) The effects of BJT self-heating on circuit behavior. *IEEE Journal of Solid-State Circuits*, **28** (6), 678–685.

[27] Boumaiza, S. and Ghannouchi, F.M. (2003) Thermal memory effects modeling and compensation in RF PAs and predistortion linearizers. *IEEE Transactions on Microwave Theory and Techniques*, **51** (12), 2427–2433.

[28] Ermolova, N.Y. (2001) Spectral analysis of nonlinear amplifier based on the complex gain Taylor series expansion. *IEEE Communications Letters*, **5** (12), 465–467.

[29] Benedetto, S., Biglieri, E. and Daffara, R. (1979) Modeling and performance evaluation of nonlinear satellite links – a Volterra series approach. *IEEE Transactions on Aerospace and Electronic Systems*, **15** (4), 494–507.

[30] Eun, C. and Powers, E.J. (1997) A new Volterra predistorter based on the indirect learning architecture. *IEEE Transactions on Signal Processing*, **45** (1), 223–228.

[31] A. Zhu and T. J. Brazil, An adaptive Volterra predistorter for the linearization of high power amplifiers. Digest IEEE International Microwave Symposium, Seattle, WA, June 2002, pp. 461–464, 2002.

[32] Zhu, A., Pedro, J.C. and Brazil, T.J. (2006) Dynamic deviation reduction based Volterra behavioral modeling of RF power amplifiers. *IEEE Transactions on Microwave Theory and Techniques*, **54** (12), 4323–4332.

[33] Crespo-Cadenas, C., Reina-Tosina, J., Madero-Ayora, M.J. and Munoz-Cruzado, J. (2010) A new approach to pruning Volterra models for RF power amplifiers. *IEEE Transactions on Signal Processing*, **58** (4), 2113–2120.

3

Model Performance Evaluation

3.1 Introduction

It is essential to accurately evaluate the performance of behavioral models and digital predistorters. This is useful for the proper selection of their structure, especially with the abundance of models that are available in the literature, as will be discussed in Chapters 4–7. Moreover, performance evaluation metrics can be adopted to decide on the model's parameters and its dimensions. These metrics can be defined either in the time or frequency domain.

This chapter is organized as follows. First, the focus will be on clearly distinguishing between the behavioral modeling and digital predistortion (DPD) applications and describing the specifics of each. Then, a variety of performance quantification metrics that have been reported in the literature for power amplifier (PA) behavioral models and digital predistorters will be thoroughly described. These are mainly categorized into two classes: time domain metrics and frequency domain metrics. Finally, the impact of memory effects on the performance assessment metrics is discussed and static nonlinearity cancelation techniques are introduced along with their relevance to behavior models and predistorter performance evaluation.

3.2 Behavioral Modeling versus Digital Predistortion

Behavioral modeling and predistortion are quite similar in various ways since most of the steps needed to derive a behavioral model or a digital predistorter are identical and most of the model structures can consistently be applied either in behavioral modeling or DPD applications. However, the performance evaluation of behavioral models is quite different from that of digital predistorters. In fact, in behavioral modeling, a structure is used to predict the output signal of the device under test (DUT) when the same input signal is applied to both the model and the DUT. Thus, as illustrated in Figure 3.1, the performance evaluation of a behavioral model is based on

Behavioral Modeling and Predistortion of Wideband Wireless Transmitters, First Edition.
Fadhel M. Ghannouchi, Oualid Hammi and Mohamed Helaoui.
© 2015 John Wiley & Sons, Ltd. Published 2015 by John Wiley & Sons, Ltd.

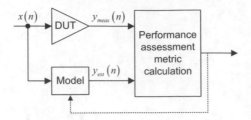

Figure 3.1   Generic block diagram for behavioral model performance assessment

comparing the measured baseband discrete time output signal samples ($y_{meas}(n)$) with those estimated by the model ($y_{est}(n)$). The similarity between these two signals can be evaluated using various metrics that are calculated either in the time domain or in the frequency domain. Figure 3.1 includes a feedback path from the performance assessment block back to the model. This is typically used to change the model structure or adjust its parameters if the model performance is not satisfactory.

In DPD applications, the performances are often evaluated using the linearity measures described in Chapter 1, namely the adjacent channel leakage ratio (ACLR) and the error vector magnitude (EVM) are calculated using the signal obtained at the output of the linearized amplifier. The AM/AM (amplitude modulation to amplitude modulation) and AM/PM (amplitude modulation to phase modulation) characteristics of the linearized DUT can also be used to assess the performance of the digital predistorter. However, this approach results in qualitative rather than quantitative estimation of the DPD performance since it only consists of visually examining the linearity of the AM/AM and AM/PM characteristics of the linearized DUT or comparing these curves to those measured on the DUT before linearization.

To evaluate digital predistorters' performances, it is also possible to consider the use of an approach similar to that employed to assess the performance of behavioral models, especially if the predistortion function is derived using the indirect learning technique in which the input and output signals of the predistorter can be derived from the measured input and output signals of the DUT. As explained in the generic block diagram of DPD systems of Figure 3.2, to ensure that the cascade made of the DUT and the digital predistorter operate as a linear amplification system, the output signal of the linearized DUT ($y_{LDUT}(n)$) should be a scaled replica of the predistorter's input signal ($x_{DPD}(n)$), the ratio being equal to the gain (G_L) of the linearized system (DPD + DUT). Accordingly, the input of the predistorter can be derived from the

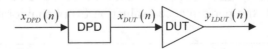

Figure 3.2   Generic block diagram of a digital predistortion system

measured output of the DUT using:

$$x_{DPD}(n) = \frac{y_{LDUT}(n)}{G_L} \tag{3.1}$$

Thus, measuring the input and output waveforms of the DUT will provide data that can be used for the training of the digital predistorter. Under such conditions, the synthesis of the digital predistorter can be perceived as a modeling problem either by considering the DPD as a standalone system as illustrated in Figure 3.3a or by considering the cascade of the DPD and DUT as displayed in Figure 3.3b. In Figure 3.3a, the performance assessment metrics are evaluated using the output of the ideal DPD and that of the DPD model to be identified. In this figure, the ideal DPD refers to a hypothetical DPD system that will generate an output signal, $y_{DPD\_ideal}(n)$, when its input signal is $x_{DPD}(n)$. Based on the scheme of Figure 3.2, the output of the ideal DPD is:

$$y_{DPD\_ideal}(n) = x_{DUT}(n) \tag{3.2}$$

Similarly, the performance of the DPD can be evaluated by considering the linearized DUT system of Figure 3.3b. In this case, the linearized DUT system is made of the DPD model and the actual DUT. Thus, the signal $y_{LDUT\_meas}(n)$ corresponds to the measured waveform at the output of the DUT when the DPD model is applied. The signal $y_{LDUT\_ideal}(n)$ represents the signal that should ideally be obtained at the output of the linearized DUT when its input signal is $x_{DPD}(n)$. Obviously, the signals $y_{LDUT\_ideal}(n)$ and $x_{DPD}(n)$ are related according to:

$$y_{LDUT\_ideal}(n) = G_L \cdot x_{DPD}(n) \tag{3.3}$$

where G_L is the linear gain of the linearized DUT.

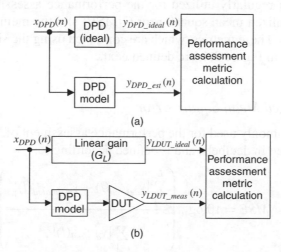

(a)

(b)

Figure 3.3 Generic block diagram for digital predistorter performance assessment (a) considering the DPD and (b) considering the linearized DUT

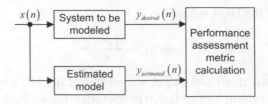

Figure 3.4 Common representation of model identification variables in time domain

In practice, evaluating the DPD performance using the approach of Figure 3.3b is more reliable since the final aim of employing the DPD is to obtain a linear system in which the input and output waveforms satisfy Equation 3.3.

The three cases described here for the definition of signals used to calculate the performance assessment metrics of behavioral models and digital predistorters can be cast in the common representation of Figure 3.4. In this figure, the input signal $x(n)$ is feeding both the systems to be modeled and its model. The output signal of the system to be modeled represents the desired signal and is labeled $y_{desired}(n)$, while the output of the model corresponds to the estimated value of the desired signal and is designated as $y_{estimated}(n)$.

3.3 Time Domain Metrics

The most straightforward approach for model performance assessment is to evaluate its prediction error that corresponds to the discrepancy between the desired and estimated output signals in the time domain. In fact, the model equation as well as the input and output signals are naturally described in a time domain. Two time domain metrics have been regularly utilized for the performance assessment of behavioral models: the normalized mean square error (NMSE) and the memory effects modeling ratio (MEMR). These metrics, which are computed using the signal $y_{desired}(n)$ and $y_{estimated}(n)$ shown in Figure 3.4, are defined next.

3.3.1 Normalized Mean Square Error

The NMSE is commonly used for the performance assessment of behavioral models. It is often expressed in decibels, and is defined according to:

$$NMSE = 10\log_{10}\left(\frac{\sum_{l=1}^{L}\left|y_{desired}(l) - y_{estimated}(l)\right|^2}{\sum_{l=1}^{L}\left|y_{desired}(l)\right|^2}\right) \tag{3.4}$$

where L refers to the length of each of the time domain waveforms $y_{desired}(n)$ and $y_{estimated}(n)$. The accuracy of a model is inversely proportional to the NMSE since a lower NMSE value indicates a superior model accuracy. Given that the power in the adjacent channels is usually much lower than that in the in-band and that the NMSE is calculated in the time domain where the contribution of these different bands is blended, this metric mainly reflects the performance of the model and its accuracy in the in-band region of the DUT output spectra [1, 2]. Thus, it is less sensitive for detecting discrepancies between the desired and estimated signals in the adjacent channels than in the in-band frequency range. Moreover, since memory effect contributions to the behavior of the DUT are much less significant than that of the static distortions, the NMSE metric applied in accordance with the setting of Figure 3.4 does not precisely expose the ability of the model to mimic the memory effects of the DUT. Thus, in behavioral modeling applications, the NMSE calculated from the signals at the output of the DUT and its model is not a reliable approach to estimating the memory depth of the system being modeled.

3.3.2   Memory Effects Modeling Ratio

The MEMR was proposed in [3], as an extension of the memory effects ratio (MER), in order to quantify the ability of a model in predicting the memory effects of the DUT. The MER measures the loss of accuracy resulting from describing a DUT having memory effects using a memoryless model. The MER was defined, in [3], as:

$$MER = 10\log_{10}\left(\frac{\sqrt{\sum_{l=1}^{L}\left|y_{desired}(l) - y_{estimated\_memoryless}(l)\right|^2}}{\sqrt{\sum_{l=1}^{L}|y_{desired}(l)|^2}}\right) \tag{3.5}$$

where L and $y_{desired}(n)$ are those defined in Equation 3.4. $y_{estimated\_memoryless}(n)$ is the predicted output waveform of the DUT using a memoryless model.

According to Equation 3.5, the MER quantifies the modeling error when the memoryless model is used. This error is made of two components. The first is attributed to the residual error in modeling the static nonlinearity of the DUT, that is, the error that would have been obtained if the DUT was memoryless. The second component of the MER is due to error caused by the presence of the memory effects that are naturally excluded from the memoryless model. Thus, the stronger the memory effects of the DUT are, the larger the MER will be.

To evaluate the accuracy enhancement achieved by increasing the memory depth of the model, it is useful to define the *mth* order error (e_m) that corresponds to the error obtained when m samples of memory are taken into account in the model. The *mth* order error is given by:

$$e_m(n) = y_{desired}(n) - y_{estimated\_mth\ order}(n) \tag{3.6}$$

with $y_{estimated\_mth\ order}(n)$ being the estimated signal at the output of the model when its memory depth is set to m.

Accordingly, the MEMR of a model with a memory depth m ($MEMR_m$) is defined as [3]:

$$MEMR_m = 1 - \frac{\sqrt{\sum_{l=1}^{L} |e_m(l)|^2}}{\sqrt{\sum_{l=1}^{L} |e_0(l)|^2}} \quad (3.7)$$

Based on Equation 3.6, the denominator of the ratio used to calculate the $MEMR_m$ in Equation 3.7 is the Euclidean norm (also known as 2-norm) or the zeroth order error (e_0), which is simply the modeling error obtained using a memoryless model. Thus, augmenting the memoryless model by adding more memory effects modeling capabilities will result in better modeling accuracy and higher MEMR. Indeed, the MEMR is lower bounded by the case where a memoryless model is used ($MEMR_0 = 0$), and upper bounded by the case where a model (including up to mth order memory effects) is perfectly reproducing the memory effects of the DUT ($MEMR_M = 1$). The MEMR can also be expressed in dB using:

$$MEMR_{m\_dB} = 10\log_{10}\left(1 - \frac{\sqrt{\sum_{l=1}^{L} |e_m(l)|^2}}{\sqrt{\sum_{l=1}^{L} |e_0(l)|^2}} \right) \quad (3.8)$$

The MEMR can be applied to estimate the appropriate memory depth of the DUT. However, the fact that static distortions prevail over the dynamic distortions renders the reliability of this approach questionable for DUTs with strong memory effects when applied directly to the measured and estimated waveforms as portrayed in Figure 3.4. The use of the MEMR metric in conjunction with static distortions cancelation techniques described in Section 3.5 can circumvent this problem.

3.4 Frequency Domain Metrics

Some performance assessment metrics are defined in the frequency domain. The main motivation is to have a more accurate estimation of the model performance in the adjacent channels since time domain signals are mainly dominated by the in-band components. Major metrics that have been applied to the performance assessment of PAs and transmitters behavioral models are described in the following subsections.

3.4.1 Frequency Domain Normalized Mean Square Error

The frequency domain NMSE ($NMSE_{FD}$) is calculated as the ratio between the power of the error signal between the desired and estimated output signals over that of the

DUT's output signal quantified over a specified bandwidth. The $NMSE_{FD}$, is expressed in dB, and is given by:

$$NMSE_{FD} = 10\log_{10}\left(\frac{\int_{f=f_{start}}^{f=f_{stop}} |E(f)|^2 df}{\int_{f=f_{start}}^{f=f_{stop}} |Y_{desired}(f)|^2 df}\right) \tag{3.9}$$

where $Y_{desired}(f)$ is the discrete time Fourier transform (DTFT) of the DUT's output signal $y_{desired}(n)$. $E(f)$ is the DTFT of the error signal defined, using the signals of Figure 3.4, as:

$$e(n) = y_{desired}(n) - y_{estimated}(n) \tag{3.10}$$

To have a comprehensive evaluation of the model performance over the entire frequency range, the integration boundaries used in Equation 3.9 should represent the limits of the output signal bandwidth. Typically if the DUT's input signal has a bandwidth BW in Hz, then the values of f_{start} and f_{stop} are commonly considered to be:

$$\begin{cases} f_{start} = -5 \times \dfrac{BW}{2} \\ f_{stop} = +5 \times \dfrac{BW}{2} \end{cases} \tag{3.11}$$

In such cases, a frequency range containing up to fifth order inter-modulation products is used as the integration bandwidth. This integration bandwidth can be adjusted based on the inter-modulation products that need to be included in the $NMSE_{FD}$ calculation. This should also take into consideration the observation bandwidth of the signal $y_{desired}(n)$ and thus the sampling rate at which this signal was acquired.

Typically, the frequency domain NMSE metric leads to results similar to that of its time domain counterpart: Though it offers an extra degree of freedom that allows for the adjustment of the integration bandwidth and consequently the inclusion or exclusion of specific frequency components.

3.4.2 Adjacent Channel Error Power Ratio

The adjacent channel error power ratio (ACEPR) was proposed in [4]. This metric makes it possible to quantify the prediction error of a model in various frequency domain ranges, for example, in the adjacent channel and/or the alternate adjacent channel. It can also be applied for the quantification of the error in the upper and lower sides of the frequency spectrum around the center frequency. The ACEPR is inspired by the adjacent channel power ratio metric and is calculated using:

$$ACEPR = 10\log_{10}\left(\frac{\int_{f=f_{start\_adjL}}^{f=f_{stop\_adjL}} |E(f)|^2 \cdot df + \int_{f=f_{start\_adjU}}^{f=f_{stop\_adjU}} |E(f)|^2 \cdot df}{\int_{f=f_{start\_channel}}^{f=f_{stop\_channel}} |Y_{desired}(f)|^2 \cdot df}\right) \tag{3.12}$$

The frequency domain signals $E(f)$ and $Y_{desired}(f)$ are those defined for Equation 3.9. $f_{start\_channel}$ and $f_{stop\_channel}$ represent the integration limits of the signal channel. Similarly, f_{start_adjL} and f_{stop_adjL} refer to the integration limits of the lower adjacent channel. f_{start_adjU} and f_{stop_adjU} are the integration limits of the upper adjacent channel.

The key difference between the $NMSE_{FD}$ of Equation 3.9 and the ACEPR of Equation 3.12 is in the integration bandwidths used to calculate the power of the error signal and that of the DUT's output signal. In fact, Equation 3.12 contrasts with Equation 3.9 since the power of the DUT's output signal and that of the error are calculated over two distinct frequency ranges. First, the power of the output signal is calculated by integrating the power of the signal $Y_{desired}(f)$ over the channel bandwidth that corresponds to the input signal's bandwidth. Moreover, the error signal's power is evaluated outside the main channel bandwidth. Indeed, in the primary definition of the ACEPR metric, the power of the error signal was calculated over the upper and lower adjacent channels using the center frequency and integration bandwidths defined in the communication standard in a manner analogous to that described in Chapter 1 for the ACLR calculation. Accordingly, this definition can be extended to employ various integration bandwidths for the calculation of the error power. This can include the estimation of the error power in the alternate adjacent channel or in both the adjacent and alternate adjacent channels. This confers to the ACEPR an enhanced flexibility to evaluate and compare model performances over specific frequency ranges. Furthermore, compared to the $NMSE_{FD}$, the ACEPR provides a more precise estimate of the model performance in the out-of-band frequency range encompassing the adjacent and alternate adjacent channels.

3.4.3 Weighted Error Spectrum Power Ratio

The weighted error spectrum power ratio (WESPR) is an extension of the ACEPR metric described previously. It was introduced first in [5] as a generalization of the ACEPR metric to any PA technology and any communication standard. As can be implied from its designation, the WESPR is calculated by applying a weighting function, in the frequency domain, to the error signal in order to control the relative importance of its frequency content. The WESPR is formulated as follows:

$$WESPR = 10\log_{10}\left(\frac{\sum_{i=1}^{I}\int_{f=f_{i\_start}}^{f=f_{i\_stop}}|W(f)\cdot E(f)|^2\cdot df}{\sum_{s=1}^{S}\int_{f=f_{s\_start}}^{f=f_{s\_stop}}|Y_{desired}(f)|^2\cdot df}\right) \qquad (3.13)$$

In this latter equation, I is the number of frequency intervals over which the power of the error signal $(E(f))$ is calculated. f_{i_start} and f_{i_stop} are the lower and upper integration limits of the ith frequency band used to calculate the error signal's power. Equivalently, S is the number of frequency bands over which the desired signal's power $(Y_{desired}(f))$

is calculated. f_{s_start} and f_{s_stop} are the lower and upper integration limits of the sth frequency band used to calculate the desired signal's power.

From Equation 3.13, it appears that the WESPR enhances the capabilities of the ACEPR by controlling the frequency bands to be used in calculating the power of the DUT's output signal and that of the error signal between the model's estimated output and the DUT's output. The WESPR is suitable for any type of signals especially multi-carrier ones where two or more active signal channels are separated by off channels. In this case, the signal and error powers can be properly estimated by adequate choice of the number of frequency bands and their corresponding power integration bandwidths.

Two weighting functions were proposed with the WESPR metric [5]. In the first function, based on hard thresholding, the weighting function magnitude is set to unity for all frequencies. The integration bandwidths for the signal power and the error power are then determined by comparing, at each frequency, the signal power to a given threshold value. The frequency bands where the power of the input signal ($x(n)$) is greater than a specified threshold value constitute the S bands used to calculate the output signal power, whereas the error power is calculated elsewhere, that is, in the frequency bands where the power of the input signal is lower than the specified threshold value.

The second weighting function is based on soft thresholding and is given by:

$$W(f) = \frac{\max(|E(f)|)}{\max(|E(f)|) + |X(f)|} \qquad (3.14)$$

where $X(f)$ is the DTFT of the DUT's input signal $x(n)$ as depicted in Figure 3.4.

Adopting the WESPR metric using the weighting function of Equation 3.14 leads to a weighted error signal where most of the energy lies in the out-of-band frequencies. Thus, this metric is suitable for accurately assessing the performance of behavioral models and their ability to mimic the behavior of the DUT in the out-of-band spectrum range.

3.4.4 Normalized Absolute Mean Spectrum Error

The three frequency domain metrics describe previously are based on the power integration of the error signal or a weighted version of it. Therefore, they all provide a macroscopic metric that characterizes the model performance. The normalized absolute mean spectrum error (NAMSE) circumvents this limitation by comparing, in the frequency domain, the spectrum of the desired output signal ($Y_{desired}(f)$) and that of the estimated output signal ($Y_{estimated}(f)$). In [1], the NAMSE was defined as:

$$NAMSE = 10\log_{10}\left[mean_{f\in[f_{min},f_{max}]}\left(\frac{|Y_{desired}(f) - Y_{estimated}(f)|}{|Y_{desired}(f)|}\right)\right] \qquad (3.15)$$

$Y_{desired}(f)$ and $Y_{estimated}(f)$ are the discrete-time Fourier transforms of the DUT's and the model's output signals ($y_{desired}(n)$ and $y_{estimated}(n)$), respectively. f_{min} and f_{max}

represent the limits of the frequency band over which the NAMSE is calculated. These limits can be customized depending on whether the NAMSE needs to be evaluated over the entire frequency range of the output signal spectrum or if it is desirable to focus it on specific frequency ranges, such as those corresponding to the adjacent channels or alternate adjacent channels.

Since it is derived from the frequency domain error, the NAMSE does not suffer from the non homogeneous power levels throughout the frequency spectrum of the signal and the significant variations of the power levels between the in-band and out-of-band frequency regions. The NAMSE gives equal weight to frequency domain errors by comparing the measured and predicted spectra without resorting to power integration. This allows for a concentrated and accurate evaluation of the model performance in the frequency domain.

3.5 Static Nonlinearity Cancelation Techniques

Behavioral models and digital predistorters currently in use inevitably take into account memory effects in addition to the static distortions. However, the contribution of memory effects to the overall behavior of the DUT is commonly much lower than that of the static distortions. Thus, it turns out that, when derived directly from the measured and estimated output waveforms ($y_{desired}(n)$ and $y_{estimated}(n)$) shown in Figure 3.4, the metrics previously reported are dominated by the static nonlinearity. They are consequently unable to accurately assess the ability of a model to predict the memory effects exhibited by the DUT [6, 7]. Therefore, in order to emphasize the memory effect modeling capabilities of a model, it is essential to cancel out the distortions due to the static nonlinearity so that the system to be modeled only includes memory effects. This can be achieved using either the static nonlinearity pre-compensation technique [8] or the static nonlinearity post-compensation technique [7]. The static nonlinearity cancelation is mostly pertinent for the behavioral modeling context. Yet, it can be applied to indirectly assess the performance of digital predistorters by evaluating their ability to accurately synthesize the desired predistorted signal. This is rarely used since predistorter performances are mostly evaluated in terms of spectrum regrowth cancelation and signal quality according to metrics such as the ACLR and EVM described in Chapter 1.

3.5.1 Static Nonlinearity Pre-Compensation Technique

The first attempt made to cancel the static nonlinearity of the DUT and that of its model was based on the pre-compensation concept [8]. This consists of applying a nonlinear function upstream of both the DUT and its model in order to eliminate their static distortions as exposed in Figure 3.5. Obviously, the nonlinear function that needs to be applied in the memoryless pre-compensator is none other than the memoryless predistorter corresponding to the measured DUT behavior.

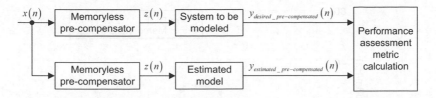

Figure 3.5   Block diagram of the static nonlinearity pre-compensation technique

To further illustrate the advantage of the static nonlinearity cancelation technique, we consider two cases of devices under test. A memoryless DUT is used in Figure 3.6, while in Figure 3.7 the DUT has memory effects. The memoryless DUT is modeled using a memoryless model. For the second DUT, two behavioral models were used to mimic its transfer function: one memoryless model and one that includes memory effects.

First, the spectrum measured at the output of this DUT as well as that predicted by its memoryless behavioral model, in accordance with the setting of Figure 3.4, were derived and are reported in Figure 3.6a. This figure shows that the model accurately estimates the output spectrum of the DUT. Then, a memoryless pre-compensator was derived and applied upstream of the DUT and its model following the scheme of Figure 3.5. Figure 3.6b presents the spectra obtained at the output of the DUT and its model after applying the memoryless pre-compensation technique. This clearly reveals the quasi-perfect similarity between the spectra of the signals at the output of the DUT and its model following memoryless pre-compensation.

The same test was repeated for a DUT exhibiting memory effects. The output signals of the DUT and its two models obtained before and after applying memoryless pre-compensation are reported in Figure 3.7a,b, respectively. One can observe, from the results of Figure 3.7a, that even though it does not take memory effects into account, the memoryless model is able to predict with a fairly good accuracy the spectrum at the output of the DUT. Indeed, by considering the frequency domain data of Figure 3.7a, it is not possible to distinguish the performances of the memoryless model from that of the model that includes memory effects. Conversely, after applying memoryless pre-compensation to the DUT and both models, one can unmistakably perceive the difference between both models' performances. In fact, canceling out the static nonlinearity allows for the evaluation of the model's ability to track the memory effects of the DUT. The spectra of Figure 3.7b demonstrate the ability of the model that includes memory effects to predict the residual nonlinearity of the DUT that is still present after applying memoryless pre-compensation. In contrast, the same figure makes evident the limitations of the memoryless model since, after memoryless pre-compensation, significant discrepancy is observed between the spectrum at the output of the DUT and that of the memoryless model.

In view of the discussion here, it is clear that static nonlinearity pre-compensation offers a valuable means of accurate assessment of behavioral models ability in

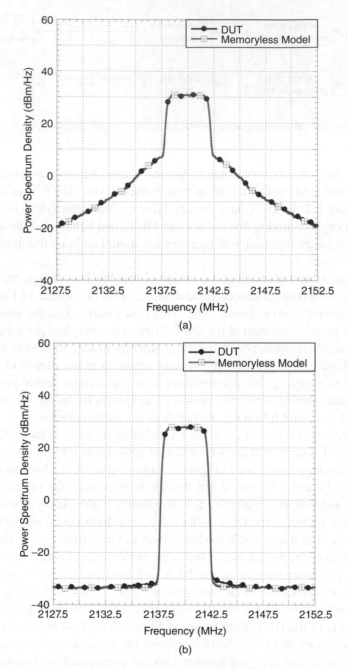

Figure 3.6 Frequency domain performance of behavioral models (case of a memoryless DUT) (a) before memoryless pre-compensation and (b) after memoryless pre-compensation

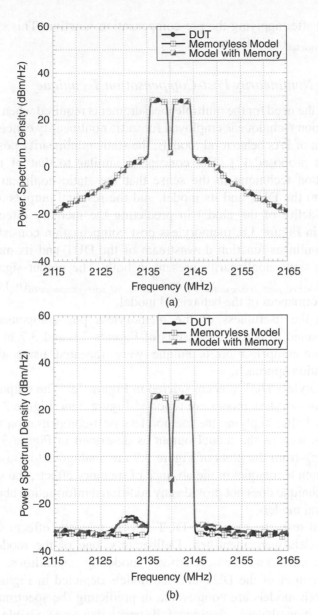

Figure 3.7 Frequency domain performance of behavioral models (case of a DUT with memory) (a) before memoryless pre-compensation and (b) after memoryless pre-compensation

predicting the memory effects of the DUT especially in the frequency domain. The main disadvantage of this technique is that the synthesis of a memoryless pre-compensation function is needed. This requires at least two sets of measurements. First, measurements are needed to derive the pre-compensation function. Then, additional measurements are needed to acquire the output signal of the system to be modeled after applying the pre-compensation function. This signal is denoted $y_{desired\_pre-compensated}(n)$ in Figure 3.5.

3.5.2 Static Nonlinearity Post-Compensation Technique

To circumvent the need for the multiple measurements required when the memoryless pre-compensation technique is employed for static nonlinearity cancelation and accurate assessment of PAs behavioral models, the static nonlinearity post-compensation technique was proposed [7]. The concept is similar to that of the memoryless pre-compensation technique in the sense that the static nonlinear distortions are eliminated from the DUT and its model, and the residual outputs are compared to evaluate the fidelity of the model in predicting the memory effects of the DUT. As illustrated in Figure 3.8, memoryless post-compensation consists of applying a memoryless nonlinear function downstream of the DUT and its model in order to cancel out the static nonlinearity present in both. The output signals obtained in each branch ($y_{desired\_post-compensated}(n)$ and $y_{estimated\_post-compensated}(n)$) are then used to assess the performances of the behavioral model.

To illustrate the usefulness of the memoryless post-compensation technique, the tests performed to derive the results of Figures 3.6 and 3.7 in the case of the memoryless pre-compensation technique were repeated using the memoryless post-compensation approach.

First, a memoryless DUT was considered in Figure 3.9. The output spectra of this DUT and its memoryless model are reported Figure 3.9a. Figure 3.9b presents the spectra obtained after applying the memoryless post-compensation technique to the DUT output as well as the model output as described in Figure 3.8. The conclusions originating from the plots of Figure 3.9 are analogous to those derived in the pre-compensation technique: in the absence of memory effects, the static distortions cancelation technique does not provide any additional information about the accuracy of the behavioral models.

In the second test, the case of a DUT having memory effects was considered. Two of its models were identified. Deliberately, one of the models was memoryless and the other had memory effects modeling capabilities. The spectra of signals at the output of the DUT and its models depicted in Figure 3.10a tend to suggest that both models are comparable in predicting the spectrum of the DUT's output signal. Nonetheless, significant disparity becomes visible after applying

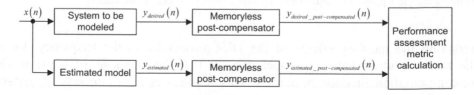

Figure 3.8 Block diagram of the static nonlinearity post-compensation technique

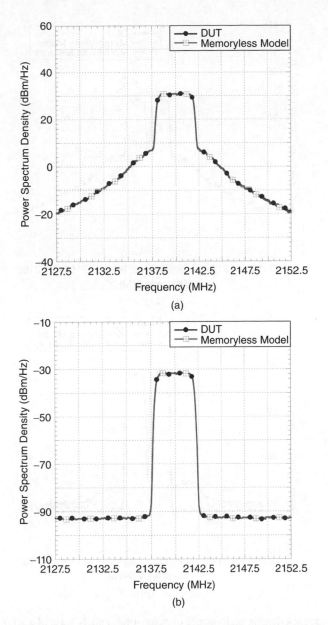

Figure 3.9 Frequency domain performance of behavioral models (case of a memoryless DUT) (a) before memoryless post-compensation and (b) after memoryless post-compensation

the memoryless post-compensation technique to the output of the DUT and each of its models as demonstrated by the spectra of Figure 3.10b. In fact, the spectra of the post-compensated memoryless model output fails to predict the residual distortions present in the post-compensated model's output. On the other hand, the post-compensated output spectrum of the model with memory effects precisely

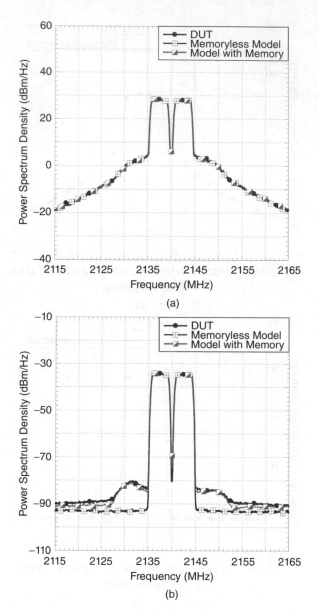

Figure 3.10   Frequency domain performance of behavioral models (case of a DUT with memory) (a) before memoryless post-compensation and (b) after memoryless post-compensation

matches the post-compensated output spectrum of the DUT. Accordingly, the memoryless post-compensation technique permits the discernment of the model performances.

The ability of the memoryless post-compensation technique in benchmarking behavioral model performances, and more specifically their ability to accurately estimate the dynamic distortions caused by a DUT, is similar to that of the pre-compensation technique. The major advantage is that the post-compensation technique requires a unique set of measurements and can be implemented in a software tool. In fact, once the waveforms at the input and output of a DUT are acquired, the behavioral model synthesis as well as the post-compensation technique can be done without requiring access to further measurements.

3.5.3 Memory Effect Intensity

The residual nonlinearity observed at the output of the post-compensator can be beneficially used to quantify the memory effects of the DUT as well as its models. This can be done using the memory effects intensity (MEI) metric [7]. The MEI is primarily used to assess the linearizability of PA prototypes by evaluating the strength of their memory effects that somehow reflect the ease of their predistortability. "Predistortability" refers to the ability of a power amplifier to be successfully linearized through digital predistortion. Yet, the MEI can also be perceived as an additional metric that can be employed to evaluate the accuracy of behavioral models by comparing its value for the post-compensated DUT output and the post-compensated model output. It is worth mentioning here that the MEI can also be derived from the output signals of a DUT and its models when the memoryless pre-compensation technique is used.

Using the signal obtained at the output of the memoryless post-compensator of Figure 3.8, the MEI is defined as the ratio between the signal power in the channel to the power of the residual dynamic distortions in the adjacent channels and/or the alternate adjacent channels [7]. Based on the sample spectrum illustrated in Figure 3.11, the MEI in the adjacent channel (MEI_{AdjCha}) and in the alternate adjacent channel ($MEI_{AltAdjCha}$) are defined by Equations 3.16 and 3.17, respectively.

$$MEI_{AdjCha} = 10\log_{10}[P_{InBand}(Y_{post-compensated})]$$
$$- 10\log_{10}[P_{AdjCha}(Y_{post-compensated})] \quad (3.16)$$

$$MEI_{AltAdjCha} = 10\log_{10}[P_{InBand}(Y_{post-compensated})]$$
$$- 10\log_{10}[P_{AltAdjCha}(Y_{post-compensated})] \quad (3.17)$$

where $P_{InBand}(Y_{post-compensated})$, $P_{AdjCha}(Y_{post-compensated})$, and $P_{AltAdjCha}(Y_{post-compensated})$ are the power of the signal $y_{post-compensated}(n)$ in the in-band, adjacent channel, and alternate adjacent channel frequency ranges respectively. $Y_{post-compensated}$ is the DTFT of the post-compensated signal $y_{post-compensated}(n)$. This post-compensated signal can be the post-compensated DUT's output signal ($y_{desired\_post-compensated}(n)$) or the post-compensated model's output signal ($y_{estimated\_post-compensated}(n)$) depending on whether the memory effects are calculated for the DUT or its model.

Without loss of generality, the in-band signal power can be expressed as:

$$P_{InBand}(Y_{post-compensated}) = \sum_{i=1}^{I} \int_{f_{ci}-\frac{BW_i}{2}}^{f_{ci}+\frac{BW_i}{2}} Y_{postcompensated}(f) \cdot df \quad (3.18)$$

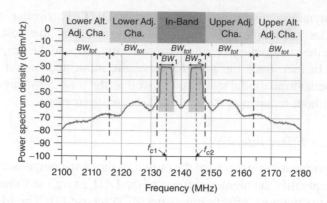

Figure 3.11   Graphical illustration of the MEI integration bandwidths

In Equation 3.18, I refers to the number of carriers in the signal $y_{post-compensated}(n)$. f_{ci} and BW_i are the center frequency and the bandwidth of ith signal carrier of $y_{post-compensated}(n)$.

The formulation of Equation 3.18 is valid for any multi-carrier signal, especially those with non-contiguous channels. In case of contiguous channels, the in-band power expression can be simplified to:

$$P_{InBand}(Y_{post-compensated}) = \int_{f_c - \frac{BW_{tot}}{2}}^{f_c + \frac{BW_{tot}}{2}} Y_{postcompensated}(f) \cdot df \qquad (3.19)$$

where f_c and BW_{tot} are the center frequency and the total bandwidth of the signal $y_{post-compensated}(n)$.

Similarly $P_{AdjCha}(Y_{post-compensated})$ and $P_{AltAdjCha}(Y_{post-compensated})$ are calculated using:

$$P_{AdjCha}(Y_{post-compensated}) = \int_{f_c - 3\frac{BW_{tot}}{2}}^{f_c - \frac{BW_{tot}}{2}} Y_{postcompensated}(f) \cdot df$$

$$+ \int_{f_c + \frac{BW_{tot}}{2}}^{f_c + 3\frac{BW_{tot}}{2}} Y_{postcompensated}(f) \cdot df \qquad (3.20)$$

$$P_{AltAdjCha}(Y_{post-compensated}) = \int_{f_c - 5\frac{BW_{tot}}{2}}^{f_c - 3\frac{BW_{tot}}{2}} Y_{postcompensated}(f) \cdot df$$

$$+ \int_{f_c + 3\frac{BW_{tot}}{2}}^{f_c + 5\frac{BW_{tot}}{2}} Y_{postcompensated}(f) \cdot df \qquad (3.21)$$

As illustrated in Figure 3.11, it possible to have significant residual distortions in the in-band frequency region of the signal at the output of the memoryless post-compensator. Specifically, these in-band inter-modulation distortions are located in the frequency range between the signal carriers. This problem becomes more pronounced as the spacing between the carriers is increased. In such conditions, accurate MEI estimation requires the inclusion of the residual power caused by the spectrum regrowth in the inter-carrier frequency range.

3.6 Discussion and Conclusion

In this chapter, key time-domain and frequency-domain metrics used for the performance assessment of PA behavioral models were discussed. These metrics provide coherent guidelines for the benchmarking of behavioral models with slightly varying resolution and diverse ability to distinguish between models with performances that might seem comparable at first glance. It was demonstrated that this ability to distinguish between models can be significantly compromised in the presence of memory effects since these are usually buried under the much stronger static distortions. Memoryless distortions cancelation through pre- or post-compensation techniques was introduced to address this limitation. Even though these alternatives for memoryless distortions cancelation are conceptually equivalent, they have specific advantages and disadvantages that were pointed out.

In conclusion, the performance assessment can be done in two steps. First the ability of a model to predict the memoryless behavior of the DUT can be evaluated using NMSE, ACEPR, or WESPR metrics with the output signals defined in the block diagram of Figure 3.4. Second, these same metrics as well as the MEMR can be applied on the output signals defined in Figure 3.5 or 3.8 to estimate the memory effects modeling aptitude of the behavioral model.

It is worth mentioning that the metrics and methods described here are not only useful to compare between various models, but they can also be applied to determine a model's dimensions and its appropriate size. Model size estimation using performance assessment metrics is an approach based strictly on approximations. Hence, the model performance typically improves as the size of the model increases. This improvement gets less and less important above a certain model size and determining the exact model size from the NMSE or any of the other metrics curves is subjective and often leads to an estimated size that is not optimal. Hybrid performance assessment metrics that quantify the model complexity and accuracy have been proposed to address the model sizing issue [9]. Furthermore, Akaike information criterion (AIC) and Bayesian information criterion (BIC) based metrics were successfully applied to select the optimal size of the model that minimizes the complexity while not jeopardizing its accuracy [10].

References

[1] Hammi, O., Younes, M. and Ghannouchi, F.M. (2010) Metrics and methods for benchmarking of RF transmitter behavioral models with application to the development of a hybrid memory polynomial model. *IEEE Transactions on Broadcasting*, **56** (3), 350–357.

[2] P. Landin, M. Isaksson, and P. Handel, Comparison of evaluation criteria for power amplifier behavioral modeling. Digest 2008 IEEE MTT-S International Microwave Symposium (IMS), Atlanta, GA, June 2008, pp. 1441–1444, 2008.

[3] Hyunchul, K. and Kenney, J.S. (2003) Behavioral modeling of nonlinear RF power amplifiers considering memory effects. *IEEE Transactions on Microwave Theory and Techniques*, **51** (12), 2495–2504.

[4] M. Isaksson, D. Wisell, and D. Ronnow , Nonlinear behavioral modeling of power amplifiers using radial-basis function neural networks. Digest 2005 IEEE MTT-S International Microwave Symposium (IMS), Long Beach, CA, June 2005, pp. 1967–1970, 2005.

[5] D. Wisell, M. Isaksson, and N. Keskitalo, A general evaluation criteria for behavioral power amplifier modeling. Proceedings 2007 69th IEEE Automatic RF Techniques Group Conference (ARFTG), Honolulu, HI, June 2007, pp. 1–5, 2007.

[6] Ghannouchi, F.M. and Hammi, O. (2009) Behavioral modeling and predistortion. *IEEE Microwave Magazine*, **10** (7), 52–64.

[7] Hammi, O., Carichner, S., Vassilakis, B. and Ghannouchi, F.M. (2008) Power amplifiers' model assessment and memory effects intensity quantification using memoryless post-compensation technique. *IEEE Transactions on Microwave Theory and Techniques*, **56** (12), 3170–3179.

[8] Liu, T., Boumaiza, S., Sesay, A.B. and Ghannouchi, F.M. (2007) Quantitative measurements of memory effects in wideband RF power amplifiers driven by modulated signals. *IEEE Microwave and Wireless Components Letters*, **17** (1), 79–81.

[9] Abdelhafiz, A.H., Hammi, O., Zerguine, A. *et al.* (2013) A PSO based memory polynomial predistorter with embedded dimension estimation. *IEEE Transactions on Broadcasting*, **59** (4), 665–673.

[10] Amiri, M.V., Bassam, S.A., Helaoui, M. and Ghannouchi, F.M. (2013) New order selection technique using information criteria applied to SISO and MIMO systems predistortion. *International Journal of Microwave and Wireless Technologies*, **5** (2), 123–131.

4

Quasi-Memoryless Behavioral Models

4.1 Introduction

Devices used in communication systems, such as traveling-wave tubes (TWTs) and solid-state power amplifiers (SSPAs), are usually described by analytical models that typically involve solving a set of simultaneous, nonlinear, partial differential equations by numerical methods. Unless some suitable simplifications can be made, such detailed models are too complex and computationally demanding to be useful in a system level simulation, where the nonlinear device is just one of many subsystems.

A higher-level model that converts an input waveform to (nearly) the correct output waveform, without necessarily resorting to the fundamental physics of the device, is needed. For linear systems, the transfer function is such a model. For nonlinear systems, the nonlinearity is represented either as a functional relationship or in tabular form for simulation applications. This representation is referred to as a *behavioral model*. It is a black-box approach to system level modeling, which provides a convenient mean of predicting system level performance without the computational complexity of full circuit model simulations.

Behavioral models can generally be divided into three groups: memoryless models, quasi-memoryless models, and behavioral models with memory. This chapter discusses how to model memoryless and quasi-memoryless nonlinear systems.

4.2 Modeling and Simulation of Memoryless/Quasi-Memoryless Nonlinear Systems

Memoryless nonlinear systems, such as power amplifiers (PAs), are usually sufficiently represented by the narrow-band AM/AM (amplitude modulation to amplitude modulation) conversion function, as no AM/PM (amplitude modulation

Behavioral Modeling and Predistortion of Wideband Wireless Transmitters, First Edition.
Fadhel M. Ghannouchi, Oualid Hammi and Mohamed Helaoui.

to phase modulation) conversion occurs in an ideal memoryless system. For a quasi-memoryless nonlinear system with a memory-time constant in the order of the period of the radio frequency (RF) carrier, the nonlinearity of the system is often represented by a set of two AM/AM and AM/PM conversion functions. Usually, AM/AM and AM/PM functions are measured by sweeping the power of an RF single tone in the center frequency of the bandpass response of the system. Class A or AB amplifiers driven by narrow-band signals around a carrier frequency are typically assumed to behave as either a memoryless nonlinear system represented by its AM/AM characteristics only, or a quasi-memoryless nonlinear system for which complex representation of both AM/AM and AM/PM characteristics is required [1–6]. The output of a memoryless/quasi-memoryless nonlinear system is a function of the input signal at the present instant only. This implies that its transfer characteristics are frequency independent.

In communication systems, the nonlinearities are usually generated and dominated by devices such as nonlinear amplifiers. Of special interest are bandpass nonlinear amplifiers used in wireless communication systems. These nonlinear devices or sub-systems are most commonly modeled by quasi-memoryless nonlinearity that exhibits nonlinear complex gain with both AM/AM and AM/PM conversions. These instantaneous representations are generally valid for bandpass signals that are considered sufficiently narrow-band, and the transfer characteristic is essentially frequency independent over the bandwidth of the signal.

Consider a narrowband signal, $\tilde{x}(t)$, with a carrier frequency at f_c:

$$\tilde{x}(t) = A(t)\cos[2\pi f_c t + \theta(t)] \qquad (4.1)$$

The characteristics of the nonlinear bandpass models are usually derived using sinusoidal wave power sweep measurements. The output of memoryless bandpass nonlinearity can be written as:

$$\tilde{y}(t) = G_A[A(t)]A(t)\cos\{2\pi f_c t + \theta(t) + \phi_G[A(t)]\} \qquad (4.2)$$

where nonlinear gain $G_A[A(t)]$ and $\phi_G[A(t)]$ are referred to as the AM/AM and the AM/PM conversion functions, respectively; and f_c is the carrier frequency.

The complex envelope signals $x(t)$ and $y(t)$ associated with the RF signals $\tilde{x}(t)$ and $\tilde{y}(t)$, respectively, are given by:

$$x(t) = A(t)e^{j\theta(t)} \qquad (4.3)$$

$$y(t) = x(t)G[A(t)] = A(t)G_A[A(t)]e^{j\{\phi_G[A(t)]+\theta(t)\}} \qquad (4.4)$$

Relations of the form given in Equation 4.2 characterize what is referred to as complex envelope nonlinearity. In an envelope nonlinearity, the nonlinear part of the output depends on the modulus $A(t)$ of the input signal only and not on its phase. A typical AM/AM and AM/PM characteristic of a TWTA (Traveling Wave Tube Amplifier) is shown in Figure 4.1.

The behavioral model of the bandpass nonlinearity as formulated in Equation 4.2 can be represented in a block diagram, as shown in Figure 4.2. Figure 4.2a is a

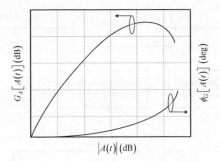

Figure 4.1 Illustration of the amplitude and phase transfer characteristics of an envelope nonlinearity of a TWTA

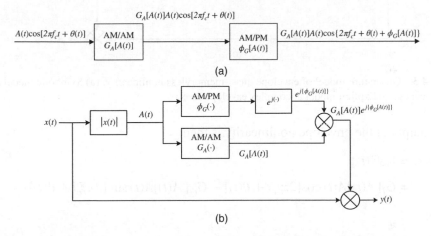

Figure 4.2 Block diagram of the behavioral model for AM/AM and AM/PM envelope nonlinearity; both $A(t)$ and $\theta(t)$ are functions of time. (a) Symbolic model at carrier frequency. (b) Explicit model at complex envelope level

high-level symbolic block diagram illustrating the conceptual flow of the model, while Figure 4.2b is a detailed block diagram explicitly showing the steps of implementing the model.

An alternative model for the same nonlinear relationship can be obtained by using the quadrature representation of bandpass systems. To demonstrate this, we can use the complex analytic representation of the signal in Equation 4.1:

$$\bar{x}(t) = A(t)e^{j[2\pi f_c t + \theta(t)]} \tag{4.5}$$

From Equation 4.5 the analytic output signal of the nonlinear memoryless system has the form of:

$$\bar{y}(t) = G_A[A(t)]A(t)e^{j\{2\pi f_c t + \theta(t) + \phi_G[A(t)]\}}$$

$$= G_A[A(t)]A(t)e^{j\{\phi_G[A(t)]\}}e^{j[2\pi f_c t + \theta(t)]}$$

$$= \{G_I[A(t)] + jG_Q[A(t)]\}\bar{x}(t) \tag{4.6}$$

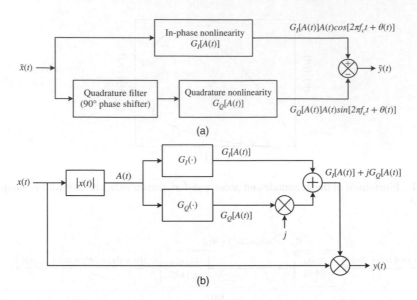

Figure 4.3 Quadrature model of envelope quasi-memoryless nonlinearity. (a) Symbolic model at carrier frequency. (b) Explicit model at complex envelope level

The output of the envelope nonlinearity is then:

$$\tilde{y}(t) = \text{Re}[\bar{y}(t)]$$
$$= G_I[A(t)]A(t)\cos[2\pi f_c t + \theta(t)] - G_Q[A(t)]A(t)\sin[2\pi f_c t + \theta(t)] \qquad (4.7)$$

where:

$$G_I[A(t)] = G_A[A(t)]\cos\{\phi_G[A(t)]\} \qquad (4.8)$$
$$G_Q[A(t)] = G_A[A(t)]\sin\{\phi_G[A(t)]\} \qquad (4.9)$$

This alternative model for bandpass quasi-memoryless nonlinearity is shown in Figure 4.3. Figure 4.3a is a high-level symbolic representation, while Figure 4.3b shows explicit steps needed for implementing the model, which are similar to what was shown in Figure 4.2. The significant point of this representation is that the tandem AM/AM and AM/PM quasi-memoryless nonlinearity can be modeled by two simple instantaneous amplitude nonlinearities, especially in communication systems where the complex signal is often in quadrature form (I and Q) instead of polar (A and θ) form in most cases.

The envelope model of bandpass nonlinearity is a very attractive form for use in simulations, as the carrier frequency, f_c, is explicit and can, therefore, be easily transformed to a low-pass equivalent model as will be demonstrated in the next section.

In this analysis, the simulator has three options with respect to the form in which $G_A(\cdot)$ and $\phi_G(\cdot)$ are expressed. The simplest method is to construct $G_A(\cdot)$ and

$\phi_G(\cdot)$ as look-up tables (LUTs) indexed with the amplitude of the input signal. This is the most commonly employed method; however, it requires relatively large memory along with appropriate interpolation techniques [7]. The second option is the use of analytical functions that best fit the measurements and the third option employs neural networks. The last two methods may introduce some inaccuracy in predicting the output of the system, but obviate the need for interpolation. These three methods are discussed further in the following sections of this chapter and Chapters 5–7.

4.3 Bandpass to Baseband Equivalent Transformation

If we consider $\tilde{x}(t)$ and $\tilde{y}(t)$ as the bandpass (or RF) input and output signals of the nonlinear system, respectively; $\tilde{x}(t)$ can be written in the form of:

$$\tilde{x}(t) = A(t)\cos[\omega_c t + \theta(t)] \tag{4.10}$$

$\tilde{x}(t)$ is considered the bandpass complex modulated signal centered on $f_c = \frac{\omega_c}{2\pi}$ with amplitude and phase modulated functions $A(t)$ and $\theta(t)$, respectively. This signal can be representative most of the digital modulation schemes. The complex bandpass representation of $\tilde{x}(t)$ can be written as:

$$\overline{x}(t) = A(t)e^{j[\omega_c t + \theta(t)]} = A(t)e^{j\omega_c t}e^{j\theta(t)} \tag{4.11}$$

With $x(t) = A(t)e^{j\theta(t)}$ the complex envelope of $\tilde{x}(t)$, such that, $\tilde{x}(t)$ is related to $\overline{x}(t)$ by:

$$\tilde{x}(t) = \mathrm{Re}[\overline{x}(t)] = \frac{\overline{x}(t) + \overline{x}^*(t)}{2} = \frac{x(t)e^{j\omega_c t} + x^*(t)e^{-j\omega_c t}}{2} \tag{4.12}$$

The RF bandpass output of the system can be related to the RF bandpass input signal via polynomial series as follows:

$$\tilde{y}(t) = f[\tilde{x}(t)] = \sum_{n=1}^{N} \tilde{a}_n \tilde{x}^n(t) = \tilde{a}_1 \tilde{x}^1(t) + \tilde{a}_2 \tilde{x}^2(t) + \cdots + \tilde{a}_N \tilde{x}^N(t) \tag{4.13}$$

Substituting Equation 4.12 into Equation 4.13, one can write:

$$\tilde{y}(t) = \sum_{n=1}^{N} \tilde{a}_n \left[\frac{x(t)e^{j\omega_c t} + x^*(t)e^{-j\omega_c t}}{2} \right]^n = \sum_{n=1}^{N} \frac{1}{2^n}\tilde{a}_n [x(t)e^{j\omega_c t} + x^*(t)e^{-j\omega_c t}]^n \tag{4.14}$$

One can expand this equation using the binomial theorem described by the following equation:

$$(a+b)^n = \sum_{k=0}^{n} a^{n-k}b^k \binom{n}{k} \tag{4.15}$$

Substituting Equation 4.14 into Equation 4.15, one can write:

$$\tilde{y}(t) = \sum_{n=1}^{N} \frac{1}{2^n} \tilde{a}_n \left\{ \sum_{k=0}^{n} \left[x(t)\, e^{j\omega_c t} \right]^{n-k} [x^*(t) e^{-j\omega_c t}]^k \binom{n}{k} \right\}$$

$$= \sum_{n=1}^{N} \frac{1}{2^n} \tilde{a}_n \left\{ \sum_{k=0}^{n} [x(t)]^{n-k} [x^*(t)]^k (e^{j\omega_c t})^{n-2k} \binom{n}{k} \right\} \tag{4.16}$$

To derive the output bandpass signal $\tilde{y}(t)$, we extract only the terms centered on ω_c from the inner summation of Equation 4.16.

Accordingly, one can obtain:

$$\tilde{y}(t) = \sum_{n=1}^{N} \frac{1}{2^n} \tilde{a}_n \{ [x(t)]^{n-k} [x^*(t)]^k (e^{j\omega_c t})^{n-2k} + [x(t)]^{n-k} [x^*(t)]^k (e^{j\omega_c t})^{n-2k} \} \binom{n}{k} \tag{4.17}$$

The relationship between $\tilde{y}(t)$ and its complex envelope $y(t)$ is:

$$\tilde{y}(t) = \frac{y(t) e^{j\omega_c t} + y^*(t) e^{-j\omega_c t}}{2} \tag{4.18}$$

Equating Equation 4.17 to Equation 4.18 term by term, one can obtain:

$$n - 2k = 1 \rightarrow \begin{cases} k = \frac{n-1}{2} \\ n - k = \frac{n+1}{2} \end{cases} \tag{4.19}$$

Based on the above Equation, n must be odd.

This leads to:

$$\tilde{y}(t) = \sum_{\substack{n=1 \\ n \text{ odd}}}^{N} \frac{1}{2^n} \tilde{a}_n \left\{ [x(t)]^{\frac{n+1}{2}} [x^*(t)]^{\frac{n-1}{2}} (e^{j\omega_c t}) + [x(t)]^{\frac{n-1}{2}} [x^*(t)]^{\frac{n+1}{2}} (e^{-j\omega_c t}) \right\} \binom{n}{\frac{n-1}{2}}$$

$$\tag{4.20}$$

Equation 4.20 can be written as:

$$\tilde{y}(t) = \sum_{\substack{n=1 \\ n \text{ odd}}}^{N} \frac{1}{2^n} \tilde{a}_n [|x(t)|^{n-1} x(t) (e^{j\omega_c t}) + |x(t)|^{n-1} x^*(t) (e^{-j\omega_c t})] \binom{n}{\frac{n-1}{2}} \tag{4.21}$$

By equating Equation 4.21 to Equation 4.18, one can obtain the output of the equivalent low-pass baseband polynomial model, $y(t)$, as a function of the complex input envelope, $x(t)$, as follows:

$$y(t) = \sum_{\substack{n=1 \\ n \text{ odd}}}^{N} \frac{1}{2^{n-1}} \binom{n}{\frac{n-1}{2}} \tilde{a}_n |x(t)|^{n-1} x(t) \tag{4.22}$$

$$y(t) = \sum_{\substack{n=1 \\ n \text{ odd}}}^{N} a_n |x(t)|^{n-1} x(t) \tag{4.23}$$

where $a_n = \frac{1}{2^{n-1}} \binom{n}{\frac{n-1}{2}} \tilde{a}_n$.

Therefore, we can directly obtain the complex envelope transfer characteristic $T(t)$, of the quasi-memoryless nonlinearity from Equation 4.23 as follows:

$$T(t) = G_I(t) + jG_Q(t) = \sum_{\substack{n=1 \\ n \text{ odd}}}^{N} a_n |x(t)|^{n-1} \tag{4.24}$$

Despite the fact that the equivalent low-pass model incorporates only odd terms, it has been demonstrated that the inclusion of even terms in the sum can help reduce the order of the polynomial needed to fit experimental data when identifying the model.

4.4 Look-Up Table Models

In various practical applications, the nonlinear characteristic is obtained experimentally; and, an adequate analytical expression may not be easily obtained. These characteristics are often depicted graphically, and the "model" for computer simulation purposes is simply a tabular representation of the experimental data. The magnitudes of the input signals and the corresponding outputs are stored in a LUT. The look-up procedure takes the input value and does a search among the discrete table entries to determine which entry is appropriate. Appropriate interpolation is used between tabulated data points, whenever necessary.

4.4.1 Uniformly Indexed Loop-Up Tables

It was mentioned earlier that quasi-memoryless nonlinearity can be represented by AM/AM and AM/PM conversion functions that depend only on the magnitude of the envelope signal, thus, there is no need to model the system using two-dimensional (2-D) LUTs due to the fact that the complex distortion is only function of the magnitude of the input signal ($\sqrt{I_{in}^2 + Q_{in}^2}$), and not the individual value of I_{in} and Q_{in}. Indeed, they can be modeled using two one-dimensional (1-D) real LUTs, one for the amplitude distortion and the second for the phase distortion. The indexing parameter of such LUT is the input signal level quantized over a finite number of levels called the depth or length of the LUT. Figure 4.4 illustrates a linear 1-D complex LUT where $x_{in}(n)$ and $x_{out}(n)$ are the LUT's input and output discrete-time signal values, respectively; $A(n)$ designates the normalized and quantized amplitude of the complex input signal and $G[A(n)]$ is the complex looked-up gain value (output) for a given quantized input signal magnitude value $A(n)$. The maximum input signal amplitude is chosen so that it corresponds to the saturation point of the nonlinear system being modeled.

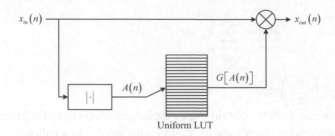

Figure 4.4   LUT for quasi-memoryless system modeling

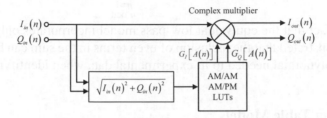

Figure 4.5   Cartesian implementation of the LUT based model for quasi-memoryless nonlinearity

$$G[A(n)] = G_I[A(n)] + jG_Q[A(n)] \qquad (4.25)$$

The LUT based model can be implemented in practice using the scheme shown in Figure 4.5 where the $I_{in}(n)$ and $Q_{in}(n)$ signals are multiplied by the complex gain value obtained from the LUT to generate the $I_{out}(n)$ and $Q_{out}(n)$ values at the output of model.

For a uniform LUT with N entries, the bin's width (step) in the LUT-index domain is equal to:

$$d = \frac{A_{\max} - A_{\min}}{N - 1} \qquad (4.26)$$

where $A_{\max}$ and $A_{\min}$ are the maximum and minimum discrete values of the quantized indexing variable A.

Synthesis of LUT entries is carried out through the implementation of Equation 4.25 in a digital processor. Each entry in the LUT is assumed to be optimal at the midpoint of its range.

4.4.2   Non-Uniformly Indexed Look-Up Tables

Since the behavior of the system is nonlinear and its complex gain distortion is function of the magnitude of the input signal, non-uniformly indexed LUTs are often used to minimize the length of the table,

$$G[A_c(n)] = G_I[A_c(n)] + jG_Q[A_c(n)] \qquad (4.27)$$

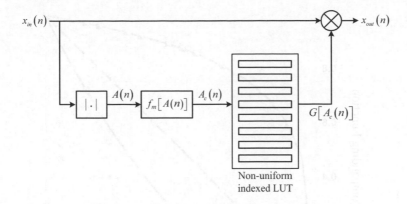

Figure 4.6 Non-uniform LUT architecture

where $G[A_c(n)]$ is the complex gain of the quasi-memoryless nonlinearity stored as entries of the LUT and indexed as function of $A_c(n)$, which is given by:

$$A_c(n) = f_m[A(n)]. \tag{4.28}$$

The companding function $f_m[A(n)]$, relating the LUT index to the input signal amplitude $A(n)$ is shown in Figure 4.6. This function allows a better distribution of the LUT entries over the whole dynamic range of the input signal, by enabling a denser distribution of LUT bins in the input signal magnitude domain whenever the amplifier gain variation changes rapidly over a given interval of the input drive level.

This function is equal to identity in the case of equally-spaced LUTs in terms of the input signal voltage magnitude. For LUTs that are equally spaced in terms of input signal power expressed in watts, the function f_m is a square function. For LUTs that are equally spaced in terms of input signal power expressed in dBm, the function f_m is a logarithmic function.

Figure 4.7 shows the values of the three companding/mapping functions (linear, square, and log) relating the index of the LUT to the normalized input signal.

The accuracy of the LUT-based models is a function of the number of entries, N, and the companding function, f_m, selected. The intermodulation distortion attributed to quantization noise generated by the LUT can be related to the derivative of the companding function, $f_m[A(n)]$, at a given input drive level and the input signal probability distribution function (pdf) as described in [8]. Therefore, for a given drive signal with a given pdf, an optimal companding function can be obtained using the methodology described in [8].

4.5 Generic Nonlinear Amplifier Behavioral Model

In a class AB amplifier, the DC power consumption, and small signal gain are functions of the input signal power level. Also, the output power expressed in decibels,

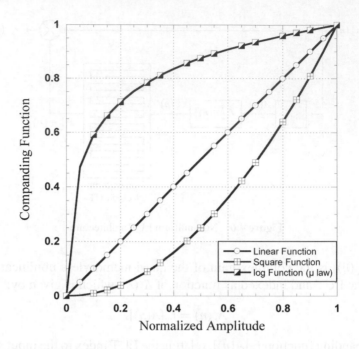

Figure 4.7 Companding functions for LUTs

P_{out}, is function of the input power, and the output phase, ϕ_{out}, is a function of the input power, all of which can be related using the following equations [9]:

$$P_{out} = P_{in} + G_{ss} - k \log_{10}\left(1 + 10^{\frac{P_{in}+G_{ss}-P_{sat}}{k}}\right) \qquad (4.29)$$

$$\phi_{out} = \phi_{in} + \mu k \log_{10}\left(1 + 10^{\frac{P_{in}+G_{ss}-P_{sat}}{k}}\right) \qquad (4.30)$$

where,

G_{ss} is the small signal gain of the amplifier expressed in decibels,
P_{sat} is the output saturated power of the amplifier expressed in decibels,
ϕ_{in} is the phase of the input signal,
k is a compression coefficient of the amplifier,
μ is the phase sensitivity to the input power level, typically it is 5 degrees per dB of compression for a single-ended amplifier.

The compression coefficient k can be estimated using the values of P_{sat} and 1 dB compression point, P_{1dB}, for a given amplifier using the following equation:

$$P_{sat} - P_{1dB} = 1 - k \log_{10}\left[10^{\left(\frac{1}{k}\right)} - 1\right] \qquad (4.31)$$

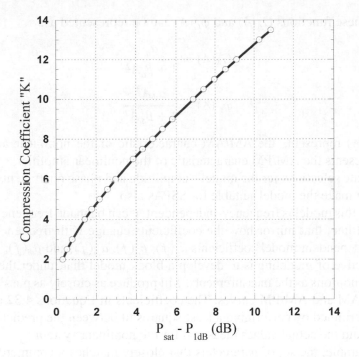

Figure 4.8 Compression coefficient k as function of P_{sat} and P_{1dB}

A direct calculation of k is not possible, but the left hand side of the equation can be estimated for different values of k, typically (1 − 10 dB) and by means of the curve obtained one can estimate the value of k for a given measured difference between P_{sat} and P_{1dB}. Figure 4.8 presents the characteristic curve of compression coefficient k calculated from Equation 4.31.

4.6   Empirical Analytical Based Models

In the remainder of this chapter and for ease of notation, the time-dependent magnitude, and phase ($A(t)$ and $\theta(t)$, respectively) of the input signal $x(t)$ will be denoted as A and θ, respectively.

4.6.1   Polar Saleh Model

In the polar representation, the output of a nonlinear amplifier can be described by the expression given by of Equation 4.2. The well-known Saleh model [10], originally developed to mimic the behavior of TWTAs, uses simple two-parameter functions to model the AM/AM and AM/PM characteristics of nonlinear quasi-memoryless systems. These functions $G_A(A)$ and $\phi_G(A)$ can be represented as:

$$G_A(A) = \frac{\alpha_a}{1 + \beta_a A^2} \tag{4.32}$$

$$\phi_G(A) = \frac{\alpha_\phi A^2}{1 + \beta_\phi A^2} \tag{4.33}$$

where $G_A(A)$ represents the AM/AM characteristic of the nonlinear amplifier and $\phi_G(A)$ represents the AM/PM characteristic of the nonlinear amplifier.

Appropriate selections for the model's amplitude and phase coefficients (α_a, β_a, α_ϕ, and β_ϕ) can make the model suitable for SSPAs also.

Although this model is frequency-independent, it can be made frequency-dependent by adding filters that mirror how the coefficients change with frequency leading to frequency-dependent model coefficients $\alpha_a(f)$, $\beta_a(f)$, $\alpha_\phi(f)$, and $\beta_\phi(f)$.

The objective of modeling is to develop a block model that, under the same input stimulus conditions as the measurements, will produce as closely as possible the measured AM/AM and AM/PM curves. The coefficients in Equations 4.32 and 4.33 are usually determined by performing a least-squares fit between the predicted values by the model and the actual values measured for the nonlinear system.

As an example, the set of parameters that closely matches a commercial TWTA's data [10] is presented in Figure 4.9. Figure 4.9 illustrates the input-output relationship based on Equations 4.32 and 4.33, and Figure 4.10 illustrates the variation of the complex gain as function of the drive level.

4.6.2   Cartesian Saleh Model

In the quadrature representation, the output of the quasi-memoryless nonlinear system is represented by Equations 4.5–4.9. The in-phase and quadrature nonlinearities, $G_I(A)$ and $G_Q(A)$, respectively, are given by:

$$G_I(A) = \frac{\alpha_I}{1 + \beta_I A^2} \tag{4.34}$$

$$G_Q(A) = \frac{\alpha_Q A^2}{(1 + \beta_Q A^2)^2} \tag{4.35}$$

As with the polar case, the coefficients in Equations 4.34 and 4.35 are also determined by a least-squares fit; and, it has been shown [10] that the resulting functions provide excellent agreement with several sets of measured data from TWTAs.

Similarly, the model can be made frequency-dependent by adding filters that mirror how the coefficients change with frequency leading to the frequency-dependent model coefficients $\alpha_I(f)$, $\beta_I(f)$, $\alpha_Q(f)$, and $\beta_Q(f)$.

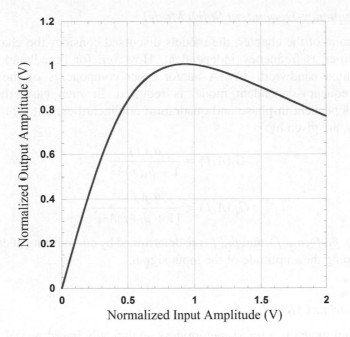

Figure 4.9 Saleh model with parameters: $\alpha_a = 2.1587$, $\beta_a = 1.1517$, $\alpha_\phi = 4.033$, and $\beta_\phi = 9.1040$

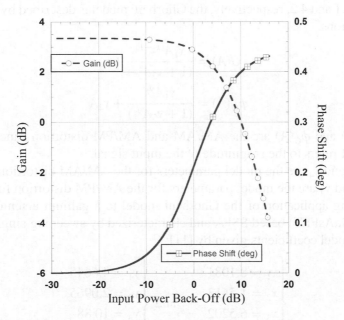

Figure 4.10 Complex gain profile for Saleh model with parameters: $\alpha_a = 2.1587$, $\beta_a = 1.1517$, $\alpha_\phi = 4.033$, and $\beta_\phi = 9.1040$

4.6.3   Frequency-Dependent Saleh Model

Until this point of the chapter, the models discussed consider the characteristics of TWT amplifier as frequency independent. However, for broadband input signals driving limited bandwidth TWTA and/or other components of the transmission chain, a frequency-dependent model is required. In such case, the power and frequency-dependent in-phase and quadrature nonlinearities, $G_I(A,f)$ and $G_Q(A,f)$, respectively, are given by:

$$G_I(A,f) = \frac{\alpha_I(f)}{1 + \beta_I(f)A^2} \tag{4.36}$$

$$G_Q(A,f) = \frac{\alpha_Q(f)A^2}{[1 + \beta_Q(f)A^2]^2} \tag{4.37}$$

where $\alpha_I(f)$, $\beta_I(f)$, $\alpha_Q(f)$, and $\beta_Q(f)$ are determined by curve fitting at each frequency while sweeping the amplitude of the input signal.

4.6.4   Ghorbani Model

The Ghorbani model is a quasi-memoryless analytically based model similar to the Saleh model [11]. In the literature, this model has been stated as being more suitable for SSPAs than the Saleh model. Considering the input and output signals given by Equations 4.1 and 4.2, respectively, the Ghorbani model is described by the following two expressions:

$$G(A) = \frac{x_1 A^{(x_2-1)}}{(1 + x_3 A^{x_2})} + x_4 \tag{4.38}$$

$$\phi_G(A) = \frac{y_1 A^{y_2}}{(1 + y_3 A^{y_2})} + y_4 A \tag{4.39}$$

where, $G(A)$ and $\phi_G(A)$ are the AM/AM and AM/PM distortion functions, respectively; and A refers to the magnitude of the input signal.

x_1, x_2, x_3, and x_4 are the model parameters for the AM/AM distortion function and y_1, y_2, y_3, and y_4 are the model parameters for the AM/PM distortion function.

A modeling application of the Ghorbani model to a gallium arsenide field-effect transistor (GaAsFET) based SSPA and characterized by sweeping single-tone power led to the model coefficients given by [11]:

$$\begin{cases} x_1 = 8.1081 \\ x_2 = 1.5413 \\ x_3 = 6.5202 \\ x_4 = -0.0718 \end{cases} \text{and} \begin{cases} y_1 = 4.6645 \\ y_2 = 2.0965 \\ y_3 = 10.88 \\ y_4 = -0.003 \end{cases}$$

For this (GaAsFET) SSPA, the Ghorbani model is shown in the Figure 4.11.

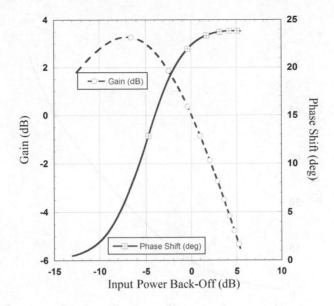

Figure 4.11 Gain and phase profiles for Ghorbani model for GaAsFET SSPA

4.6.5 Berman and Mahle Phase Model

In [12], Berman and Mahle proposed a model suited to TWTAs for multiple access communications satellite applications. This model only represents the AM/PM distortions that are given by the following form:

$$\phi_G(A) = k_1(1 - e^{-k_2 A^2}) + k_3 A^2 \qquad (4.40)$$

where A refers to the magnitude of the input signal and $\phi_G(A)$ is phase shift, relative to the output phase in small signal conditions, introduced to the output signal by the AM/PM distortion. The values for model coefficients k_1, k_2, and k_3 are found through optimization. Figure 4.12 presents the relative phase shift versus the normalized input power characteristic calculated with Equation 4.40 for a TWTA. The values of the coefficients used to generate Figure 4.12 are $k_1 = 0.372$, $k_2 = 5.14$, $k_3 = 0.27$ [12].

4.6.6 Thomas–Weidner–Durrani Amplitude Model

Thomas, Weidner, and Durrani presented a method in [13] to model the normalized amplitude's nonlinearity in TWTAs. The mathematical expression for the model can be given as:

$$M(A) = 10^{\alpha\left\{\cos\left[\frac{\log_{10}\left(\frac{A}{A_s}\right)}{\beta}\right]-1\right\}} \quad \text{for } A > A_c \qquad (4.41)$$
$$M(A) = A \qquad \qquad \qquad \text{for } A \leq A_c$$

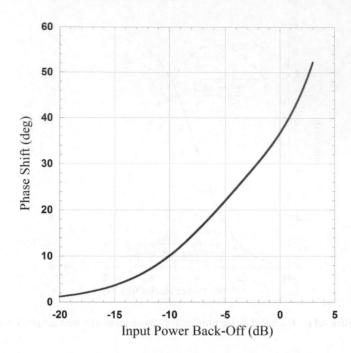

Figure 4.12   Phase shift transfer function characteristic of the Berman and Mahle model

where, here also A refers to the magnitude of the input signal, $M(A)$ is the level of the small-signal normalized output signal, A_s is the input saturation level, and A_c is the input signal's level where the compression starts. Model coefficients α and β are normally found by optimization to fit experimental data. In this model, the phase distortion is modeled using the Berman and Mahle model given by Equation 4.40.

Figure 4.13 shows the AM/AM and AM/PM characteristics for TWTA, where AM/AM characteristic is obtained from Thomas–Weidnar–Durrani model and the AM/PM characteristic is approximated using the Berman–Mahle model.

4.6.7 Limiter Model

The ideal limiter (clipping) model AM/AM relationship can be expressed as:

$$y(t) = \begin{cases} \dfrac{y_{sat}}{x_{sat}}x(t) & for \ |x(t)| < x_{sat} \\ y_{sat} & for \ |x(t)| \geq x_{sat} \end{cases} \qquad (4.42)$$

where $x(t)$ and $y(t)$ are the baseband input and output signals, respectively; x_{sat} and y_{sat} are the input and output saturation (clipping) levels, respectively.

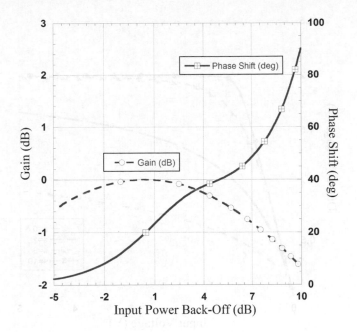

Figure 4.13 AM/AM and AM/PM characteristics for TWTA modeled using the Thomas–Weidner–Durrani model

The general limiter model baseband input-output relationship can be used for resistive memoryless nonlinearity that has no phase distortion and it can be described by:

$$y(t) = \frac{y_{sat}}{\left\{1 + \left[\frac{x_{sat}}{|x(t)|}\right]^{s}\right\}^{\frac{1}{s}}} x(t) \qquad (4.43)$$

where y_{sat} is the output saturation value, x_{sat} is the input saturation value and s is the compression shaping parameter. Note that $s \approx 0$ corresponds to a soft limiter and $s \approx \infty$ corresponds to a hard limiter. Figure 4.14 shows the limiter characteristics for different values of the parameter s.

4.6.8 ARCTAN Model

The general form of the ARCTAN (arctangent) bandpass input-output relationship can be used for memoryless nonlinearity and it is described by:

$$y(t) = \{\gamma_1 \tan^{-1}[\alpha_1 A] + \gamma_2 \tan^{-1}[\alpha_2 A]\} e^{j\theta} \qquad (4.44)$$

where A and θ are the amplitude and phase of the input signal $x(t)$ as defined in Equation 4.1, respectively. γ_1, γ_2, α_1, and α_2 are the model coefficients. Figure 4.15

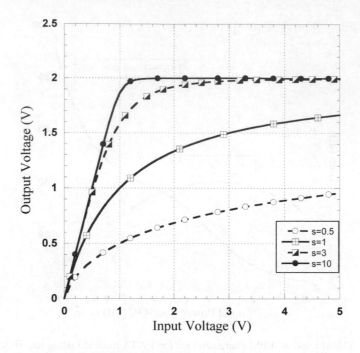

Figure 4.14 Limiter characteristics

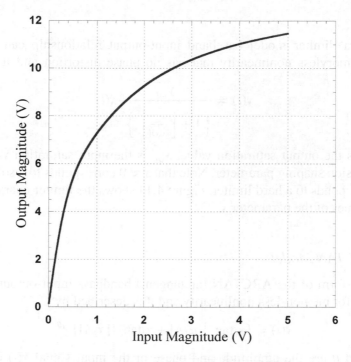

Figure 4.15 ARCTAN model characteristics

depicts the AM/AM characteristics of the ARCTAN model for $\gamma_1 = 8.0035 - j4.6116$, $\gamma_2 = -3.7717 + j12.0376$, $\alpha_1 = 2.2689$, and $\alpha_2 = 0.8234$.

4.6.9 Rapp Model

The general form of the Rapp input-output relationship can be used for memoryless nonlinearity and it is described by [14]:

$$y(t) = \frac{G_{ss}}{\left[1 + \left|\frac{x(t)}{x_{sat}}\right|^{2\sigma}\right]^{\frac{1}{2\sigma}}} x(t) \tag{4.45}$$

where $x(t)$ and $y(t)$ are the model's input and output signals, respectively. σ is a positive smoothing factor, x_{sat} is the input saturation value, and G_{ss} is the small-signal gain. Note than when $\sigma \approx 0$, the model approaches the soft limiter's behavior; and when $\sigma \approx \infty$, it approaches the hard limiter behavior.

This model was found to be suitable for SSPAs. Figure 4.16 shows the characteristics of the Rapp model for different values of σ.

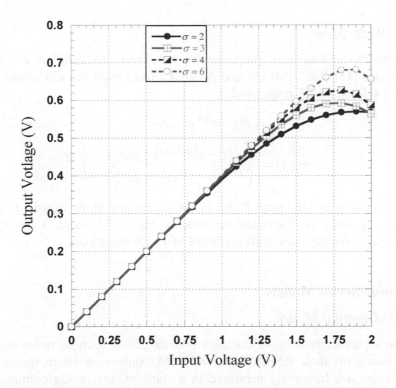

Figure 4.16   Rapp model characteristics for different smoothing factor σ

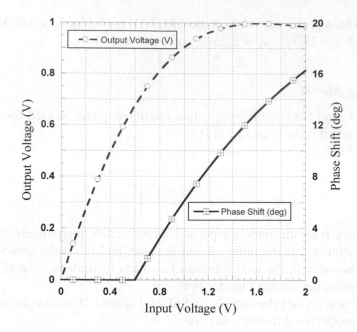

Figure 4.17   AM/AM and AM/PM characteristics of the White model

4.6.10   White Model

The White model has been proposed for baseband modeling of complex gain nonlinearities [15]. The AM/AM and AM/PM functions of the model are shown in Equations 4.46 and 4.47, respectively:

$$|y(t)| = a(1 - e^{-bA}) + cAe^{-dA^2} \tag{4.46}$$

$$\phi_G(A) = \begin{cases} f\left[1 - e^{-g(A-h)}\right] & for \ A \geq h \\ 0 & for \ A < h \end{cases} \tag{4.47}$$

where $|y(t)|$ and A are the magnitudes of the output and input signals, respectively. a, b, c, d, f, g, and h are the model coefficients. Figure 4.17 depicts a sample of the output voltage and the phase shift predicted by the White model.

4.7   Power Series Models

4.7.1   Polynomial Model

A common characterization of complex nonlinearities, such as quasi-memoryless PAs, is based on their AM/AM and AM/PM conversion characteristics. These characteristics are frequently measured in a static manner using continuous wave (CW) signals and network analyzer based setups. However, a more accurate PA characterization can be achieved by performing dynamic AM/AM and AM/PM measurements using modulated signals and devoted setups as it will be discussed in Chapter 8 [16–18]. To predict the spectral regrowth of such PAs, the complex envelopes of the RF input and output signals can be related by:

$$y(t) = x(t)G(A) \tag{4.48}$$

where $A = |x(t)|$ and $G(A)$ is the complex gain of the PA defined as:

$$G(A) = G_A(A)e^{j\phi_G(A)} \tag{4.49}$$

where $G_A(A)$ is the magnitude of the complex gain $G(A)$ and corresponds to the AM/AM conversion. The phase of the complex gain $G(A)$, $\phi_G(A)$, represents the output phase shift and corresponds to the AM/PM conversion. The complex envelope nonlinearity $G(A)$ can be represented by a complex polynomial power series of a finite order N such that:

$$y(t) = \sum_{k=1}^{N} a_k |x(t)|^{k-1} x(t) = \sum_{k=1}^{N} a_k \Psi_k^P[x(t)] \tag{4.50}$$

where $\Psi_k^P[x(t)] = |x(t)|^{k-1} x(t)$ are the basis functions of the polynomial model, and a_k are the model's complex coefficients. Several polynomial based models have been proposed for the modeling and predistortion of nonlinear PAs and transmitters. These are thoroughly discussed in Chapter 5.

4.7.2 Bessel Function Based Model

Bessel function series approximation is employed to model the amplitude and phase characteristics of memoryless nonlinear devices. It was found to be suitable for TWTAs. The approximating expression is given by:

$$G = G_I + jG_Q = \sum_{k=1}^{N} a_k J_1(\alpha k A) \tag{4.51}$$

where G_I and G_Q are the real and imaginary components of the actual measured single unmodulated carrier envelope amplitude and phase transfer characteristics. In Equation 4.51, a_k are complex coefficients, $J_1(\cdot)$ is a Bessel function of the first kind [19]. α is an arbitrary constant for scaling the input signal level depending on the over drive of the nonlinear system under consideration, which can be calculated by:

$$\alpha = 2\pi 10^{-\frac{X}{20}} \tag{4.52}$$

where X is the overdrive power level in decibels.

$J_1(\alpha k A)$ is a Bessel function of the first kind, which is defined by:

$$J_1(\alpha k A) = \sum_{m=0}^{1} \frac{(-1)^m}{m!(m+1)!}\left(\frac{\alpha k A}{2}\right)^{2m+1} \tag{4.53}$$

The output of a Bessel series baseband complex envelope based behavioral model can be expressed by the following equation:

$$y_{BP}(t) = \sum_{k=1}^{N} a_k J_1(\alpha k A)x(t) = \sum_{k=1}^{N} a_k \Psi_k^{B1}[x(t)] \tag{4.54}$$

where $y_{BP}(t)$ is the complex envelope output using the Bessel polynomials, a_k are the complex coefficients of the model, $x(t)$ is the baseband input signal, and A its magnitude; and $\Psi_k^{B1}[x(t)] = J_1(\alpha k A)x(t)$ are the basis functions of the Bessel function of the first kind based model.

4.7.3 Chebyshev Series Based Model

The output of the Chebyshev series complex envelope based model is described by following equations:

$$G = \sum_{k=0}^{N} a_k T_k(A)^{k-1} \tag{4.55}$$

$$y_{CP}(t) = \sum_{k=0}^{N} a_k T_k(A)^{k-1}x(t) = \sum_{k=0}^{N} a_k \Psi_k^{C}[x(t)] \tag{4.56}$$

where, $\Psi_k^{C}[x(t)]$ are basis functions of the Chebyshev series based model. a_k are the complex coefficients of the model to be identified, $x(t)$ is the baseband input signal, and $T_k(A)$ are the Chebyshev functions defined by the following equation:

$$T_k(A) = \cos\{k[\cos^{-1}(A)]\} \tag{4.57}$$

where $T_0(A) = 1$; $T_1(A) = A$; $T_2(A) = 2AT_1(A) - T_0(A)$; $T_3(A) = 2AT_2(A) - T_1(A)$, and so on.

The basis functions, $T_k(A)$, used in the model here are orthogonal in the interval $[-1, 1]$. A modified set of Chebyshev functions, $T_k(A^2)$, has also been used in the behavior modeling of nonlinear systems.

4.7.4 Gegenbauer Polynomials Based Model

The Gegenbauer complex polynomials were proposed in [20]. The output of this model is related to the input through:

$$y_{GP}(t) = G(A)x(t) = \sum_{k=1}^{N} a_k \Psi_k^{G}[x(t)] \tag{4.58}$$

Table 4.1   First five Gegenbauer polynomial basis functions

$$C_0^\lambda(A) = 1$$

$$C_1^\lambda(A) = 2\lambda A$$

$$C_2^\lambda(A) = 2\lambda(\lambda + 1)A^2 - \lambda$$

$$C_3^\lambda(A) = \frac{4}{3}\lambda(\lambda + 1)(\lambda + 2)A^3 - 2\lambda(\lambda + 1)A$$

$$C_4^\lambda(A) = \frac{2}{3}\lambda(\lambda + 1)(\lambda + 2)(\lambda + 3)A^4 - 2\lambda(\lambda + 1)(\lambda + 2)A^2 - \frac{1}{2}\lambda(\lambda + 1)$$

where, a_k are the model's complex coefficients, and $\Psi_k^G[x(t)]$ are the basis functions expressed with Gegenbauer polynomials according to:

$$\Psi_k^G[x(t)] = x(t)C_{k-1}^\lambda(A) \tag{4.59}$$

where λ is a signal dependent parameter that takes values in the interval $[0.8, 0.9]$ for typical wireless signals.

For $k \geq 2$, the recurrence relation of Gegenbauer polynomials of the basis function $\Psi_k^G[x(t)]$ can be expressed as:

$$C_k^\lambda(A) = \frac{2}{k}(k + \lambda - 1)AC_{k-1}^\lambda(A) - \frac{1}{k}(k + 2\lambda - 2)C_{k-2}^\lambda(A) \tag{4.60}$$

where $C_0^\lambda(A) = 1$ and $C_1^\lambda(A) = 2\lambda A$.

The Gegenbauer polynomials can be made orthogonal over the internal $[-1, 1]$ by applying the terms $C_k^\lambda(\cdot)$ on A_0 rather than A, where A_0 is given by:

$$A_0 = \frac{2A - \beta - \alpha}{\beta - \alpha} \tag{4.61}$$

with α and β defined such that $A \in [\alpha, \beta]$.

Table 4.1 shows the first five Gegenbauer polynomials basis functions $C_k^\lambda(A)$.

4.7.5   Zernike Polynomials Based Model

The Zernike polynomials based envelope nonlinearity model was proposed in [21]. The output of this polynomial model is given by:

$$y_{ZP}(t) = G(A)x(t) = \sum_{k=1}^{N} a_k \Psi_k^Z[x(t)] \tag{4.62}$$

where, a_k are the model's complex coefficients, $x(t)$ is the input signal and $\Psi_k^Z[x(t)]$ are the Zernike polynomials basis functions given by:

$$\Psi_k^Z[x(t)] = \sum_{l=0}^{\alpha} U_{(2l+\delta)k}|x(t)|^{2l+\delta}x(t) \tag{4.63}$$

Table 4.2   First nine basis functions $\Psi_k^Z[x(t)]$

k	$\Psi_k^Z[x(t)]$
0	$x(t)$
1	$x(t)\lvert x(t)\rvert$
2	$2x(t)\lvert x(t)\rvert^2 - x(t)$
3	$3x(t)\lvert x(t)\rvert^3 - 2x(t)\lvert x(t)\rvert$
4	$6x(t)\lvert x(t)\rvert^4 - 6x(t)\lvert x(t)\rvert^2 + x(t)$
5	$10x(t)\lvert x(t)\rvert^5 - 12x(t)\lvert x(t)\rvert^3 + 3x(t)\lvert x(t)\rvert$
6	$20x(t)\lvert x(t)\rvert^6 - 30x(t)\lvert x(t)\rvert^4 + 12x(t)\lvert x(t)\rvert^2 - x(t)$
7	$35x(t)\lvert x(t)\rvert^7 - 60x(t)\lvert x(t)\rvert^5 + 30x(t)\lvert x(t)\rvert^3 - 4x(t)\lvert x(t)\rvert$
8	$70x(t)\lvert x(t)\rvert^8 - 140x(t)\lvert x(t)\rvert^6 + 90x(t)\lvert x(t)\rvert^4 - 20x(t)\lvert x(t)\rvert^2 + x(t)$

where $\alpha = floor\left(\frac{k}{2}\right)$ and $\delta = \mathrm{mod}(k, 2)$ with,

$$
U_{(2l+\delta)k} = \begin{cases} (-1)^{\alpha-l}\dfrac{(\alpha+l+\delta)!}{(\alpha-l)!\,l!\,(l+\delta)!} & for \;\; (2l+\delta) \leq k \\ 0 & otherwise \end{cases}. \tag{4.64}
$$

Table 4.2 shows the first nine basis functions $\Psi_k^Z[x(t)]$.

References

[1] Gard, K.G., Gutierrez, H.M. and Steer, M.B. (1999) Characterization of spectral regrowth in microwave amplifiers based on the nonlinear transformation of a complex Gaussian process. *IEEE Transactions on Microwave Theory and Techniques*, **47** (7), 1059–1069.

[2] H. Gutierrez, K. Gard, and M. B. Steer, Spectral regrowth in microwave amplifiers using transformation of signal statistics. Digest 1999 IEEE MTT-S International Microwave Symposium (IMS), Anaheim, CA, June 1999, pp. 985–988, 1999.

[3] Wu, Q., Xiao, H. and Li, F. (1998) Linear RF power amplifier design for CDMA signals: a spectrum analysis approach. *Microwave Journal*, **41** (12), 22–40.

[4] Clark, C.J., Chrisikos, G., Muha, M.S. *et al.* (1998) Time-domain envelope measurement technique with application to wideband power amplifier modeling. *IEEE Transactions on Microwave Theory and Techniques*, **46** (12), 2531–2540.

[5] A. Leke and J.S. Kenny, Behavioral modeling of narrowband microwave power amplifiers with applications in simulating spectral regrowth. Digest 1999 IEEE MTT-S International Microwave Symposium (IMS), San Francisco, CA, June 1996, pp. 1385–1388, 1996.

[6] Chen, S., Panton, W. and Gilmore, R. (1996) Effects of nonlinear distortion on CDMA communication systems. *IEEE Transactions on Microwave Theory and Techniques*, **44** (12), 2743–2750.

[7] Meijering, E. (2002) A chronology of interpolation: from ancient astronomy to modern signal and image processing. *Proceedings of the IEEE*, **90** (3), 319–342.

[8] Boumaiza, S., Li, J., Jaidane-Saidane, M. and Ghannouchi, F.M. (2004) Adaptive digital/RF pre-distortion using a nonuniform LUT indexing function with built-in dependence on the amplifier nonlinearity. *IEEE Transactions on Microwave Theory and Techniques*, **52** (12), 2670–2677.

[9] Turlington, T.R. (2000) *Behavioral Modeling of Nonlinear RF and Microwave Devices*, Artech House, Boston.

[10] Saleh, A.A.M. (1981) Frequency-independent and frequency-dependent nonlinear models of TWT amplifiers. *IEEE Transactions on Communications*, **29** (11), 1715–1720.

[11] A. Ghorbani, and M. Sheikhan, The effect of solid state power amplifiers (SSPAs) nonlinearities on MPSK and M-QAM signal transmission. Proceedings Sixth International Conference on Digital Processing of Signals in Communications, Loughborough, UK, September 1991, pp. 193–197, 1991.

[12] Berman, A. and Mahle, C.E. (1970) Nonlinear phase shift in traveling-wave tubes as applied to multiple access communications satellites. *IEEE Transactions on Communication Technology*, **18** (1), 37–48.

[13] Thomas, C., Weidner, M. and Durrani, S. (1974) Digital amplitude-phase keying with M-array alphabets. *IEEE Transactions on Communications*, **22** (2), 168–180.

[14] C. Rapp, Effects of HPA-nonlinearity on a 4-DPSK/OFDM signal for a digital sound broadcasting system. Proceedings Second European Conference on Satellite Communications, Liege, Belgium, October 1991, pp. 179–184, 1991.

[15] White, G.P., Burr, A.G. and Javornik, T. (2003) Modeling of nonlinear distortion in broadband fixed wireless access systems. *Electronics Letters*, **39** (8), 686–687.

[16] Hammi, O., Younes, M. and Ghannouchi, F.M. (2010) Metrics and methods for benchmarking of RF transmitter behavioral models with application to the development of a novel hybrid memory polynomial model. *IEEE Transactions on Broadcasting*, **56** (3), 350–357.

[17] Jebali, C., Boulejfen, N., Gharsallah, A. and Ghannouchi, F.M. (2012) Effects of signal PDF on the identification of behavioral polynomial models for multicarrier RF power amplifiers. *Analog Integrated Circuits and Signal Processing*, **73** (1), 217–224.

[18] Schreurs, D., O'Droma, M., Goacher, A.A. and Gadringer, M. (2009) *RF Power Amplifier Behavioral Modeling*, Artech House, Boston, MA.

[19] O'Droma, M.S. (1989) Dynamic range and other fundamentals of the complex Bessel function series approximation model for memoryless nonlinear devices. *IEEE Transactions on Communications*, **37** (4), 397–398.

[20] Harguem, A., Boulejfen, N., Ghannouchi, F.M. and Gharsallah, A. (2014) Robust behavioral modeling of dynamic nonlinearities using Gegenbauer polynomials with application to RF power amplifiers. *International Journal of RF and Microwave Computer-Aided Engineering*, **24** (2), 268–279.

[21] Jebali, C., Boulejfen, N., Rawat, M. *et al.* (2012) Modeling of wideband radio frequency power amplifiers using Zernike polynomials. *International Journal of RF and Microwave Computer-Aided Engineering*, **22** (3), 289–296.

5

Memory Polynomial Based Models

5.1 Introduction

Radio frequency (RF) power amplifiers and transmitters are considered to be dynamic nonlinear systems that concurrently exhibit static nonlinearities as well as nonlinear memory effects. Thus, the most comprehensive behavioral model that can be adopted to fully model such systems is the Volterra series described in Chapter 2. However, full Volterra series (which do not include any simplifying assumptions) are typically difficult to manipulate and result in unrealistically large models that are not suitable in practice.

The memory polynomial model represents a very compact version of the Volterra series and has been widely applied in the behavioral modeling and predistortion of power amplifiers and transmitters having memory effects. A wide assortment of structures based on the memory polynomial has been proposed for the modeling and predistortion of RF power amplifiers and transmitters. Although the functions reported in this chapter are commonly referred to as models, they can be seamlessly used in both behavioral modeling and digital predistortion applications.

Figure 5.1 illustrates the two methodologies that can be recognized as the rationals behind the development of a large number of memory polynomial based structures. As shown in this figure, the memory polynomial model has low complexity but relatively limited performance; whereas the Volterra series model has a higher complexity but leads to better performance. Thus, the first approach in the development of memory polynomial based models is aimed at reducing the number of coefficients of the Volterra series, while maintaining satisfactory accuracy. Conversely, the second approach consists in augmenting the memory polynomial model to improve its performance with minimal increase in the model complexity. The goal is an ideal model that combines both high performance and low complexity.

Behavioral Modeling and Predistortion of Wideband Wireless Transmitters, First Edition.
Fadhel M. Ghannouchi, Oualid Hammi and Mohamed Helaoui.
© 2015 John Wiley & Sons, Ltd. Published 2015 by John Wiley & Sons, Ltd.

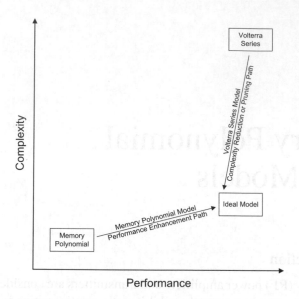

Figure 5.1   Trends in single-box memory polynomial models development

5.2   Generic Memory Polynomial Model Formulation

Memory polynomial based models are commonly used as standalone single-box models. However, they can also be part of two-box structures or in general multi-box structures. In this chapter, the main focus is memory polynomial based single-box models. Their implementation in two-box structures will be discussed in Chapter 6.

At this point, it is important to distinguish between the concept of a single-box model and a single-basis function model. A single-box model refers to a structure that is identified as a single function, in contrasts with a multi-box model where the sub-functions are determined successively. Thus, a single-box model can have more than one single basis function as, for example, is the case in the generalized memory polynomial model. This concept is further discussed in the subsequent sections of this chapter where single-box multi-basis function models are described and their mathematical formulations are derived.

All memory polynomial models can be formulated using the same generic linear system given by:

$$y(n) = \boldsymbol{\varphi}(n) \cdot \mathbf{A} \tag{5.1}$$

where $y(n)$ is the model's baseband complex output sample at instant n, $\boldsymbol{\varphi}(n)$ is a vector built using the baseband complex input signal samples according to the model's basis functions set, and $\mathbf{A}$ is the vector containing the model coefficients.

The formulation of Equation 5.1 is valid for all memory polynomial models, independent of their type and the number of basis functions. The only difference is that vector $\boldsymbol{\varphi}(n)$ is defined depending on the model. It is important to notice here that all memory polynomial models are linear with respect to their coefficients, which enables the use of simple identification techniques.

5.3 Memory Polynomial Model

Kim and Konstantinou proposed the memory polynomial model [1]. It can be obtained by reducing the Volterra series model to its diagonal terms, that is, by removing all cross-terms. The baseband complex output signal ($\mathbf{y_{MP}}$) of the memory polynomial model is expressed as a function of its baseband complex input signal ($\mathbf{x}$) according to:

$$y_{MP}(n) = \sum_{m=0}^{M} \sum_{k=1}^{K} a_{mk} \cdot x(n-m) \cdot |x(n-m)|^{k-1} \tag{5.2}$$

where a_{mk} represent the model's coefficients; and K and M are the model's nonlinearity order and memory depth, respectively.

Equation 5.2 can be rewritten in the generic formulation of Equation 5.1

$$y_{MP}(n) = \boldsymbol{\varphi}_{MP}(n) \cdot \mathbf{A} \tag{5.3}$$

where $\boldsymbol{\varphi}_{MP}(n)$ and $\mathbf{A}$ are defined by:

$$\boldsymbol{\varphi}_{MP}(n) = \begin{bmatrix} x(n) \\ \vdots \\ x(n) \cdot |x(n)|^{K-1} \\ x(n-1) \\ \vdots \\ x(n-1) \cdot |x(n-1)|^{K-1} \\ \vdots \\ x(n-M) \cdot |x(n-M)|^{K-1} \end{bmatrix}^{T} \tag{5.4}$$

$$\mathbf{A} = \begin{bmatrix} a_{01} & \cdots & a_{0K} & a_{11} & \cdots & a_{1K} & \cdots & a_{MK} \end{bmatrix}^{T} \tag{5.5}$$

where $[\cdot]^{T}$ denotes the transpose operator. Based on Equation 5.2, the memory polynomial model has two degrees of freedom since its dimension is defined by both the nonlinearity order and the memory depth.

The block diagram of the memory polynomial model is shown in Figure 5.2. This figure shows that the model can be seen as the combination of $(M + 1)$ polynomial functions, each of which is applied to a delayed version of the baseband complex input sample, $x(n)$.

5.4 Variants of the Memory Polynomial Model

5.4.1 Orthogonal Memory Polynomial Model

One of the major drawbacks of the memory polynomial model is the ill-conditioning of its data matrix that needs to be constructed when identifying the model coefficients. To better illustrate this aspect, let's rewrite Equation 5.3 for L samples (where L

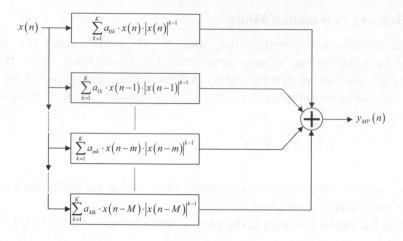

Figure 5.2   Block diagram of the memory polynomial model

refers to the length of the data set used to identify the memory polynomial model coefficients):

$$\begin{bmatrix} y_{MP}(n) \\ y_{MP}(n-1) \\ \vdots \\ y_{MP}(n-L+1) \end{bmatrix} = \mathbf{\Theta}_{MP}(n,L) \cdot \mathbf{A} \tag{5.6}$$

where the data matrix $\mathbf{\Theta}_{MP}(n,L)$ is given by:

$$\mathbf{\Theta}_{MP}(n,L) = \begin{bmatrix} \boldsymbol{\varphi}_{MP}(n) \\ \boldsymbol{\varphi}_{MP}(n-1) \\ \vdots \\ \boldsymbol{\varphi}_{MP}(n-L+1) \end{bmatrix} \tag{5.7}$$

where $\boldsymbol{\varphi}_{MP}(n)$ is as defined in Equation 5.4.

The nature of the memory polynomial model introduces a significant amount of correlation in the data matrix, $\mathbf{\Theta}_{MP}(n,L)$, especially between the elements of each row ($\boldsymbol{\varphi}_{MP}$). Moreover, a lesser degree of correlation is also present between the various rows of $\mathbf{\Theta}_{MP}(n,L)$, due to the inherent correlation of the signal's consecutive samples. The correlation of the data matrix results in an ill-conditioning problem that makes the linear system of Equation 5.6 vulnerable to disturbances [2] and numerical instability [3, 4]. This problem gets even more pronounced as the nonlinearity order of the model is increased.

To alleviate the ill-conditioning problem described here, the orthogonal memory polynomial model was proposed [3]. In this model, a new set of orthogonal basis functions is used. The basis functions are derived for signals whose magnitudes are uniformly distributed in the of range [0, 1]. When used with standard compliant communication signals that have a different distribution function such as Raleigh distribution, the advantage of the orthogonal memory polynomial model in terms of condition number reduction, is significantly decreased. However, it still compares favorably with the memory polynomial model, as it achieves moderate condition number reduction for comparable performance and implementation complexity.

The output (y_{OMP}) of the orthogonal memory polynomial model is related to its input (x) according to:

$$y_{OMP}(n) = \sum_{m=0}^{M} \sum_{k=1}^{K} a_{mk} \cdot \psi_k[x(n-m)] \qquad (5.8)$$

where a_{mk}, K, and M are as defined for Equation 5.2, and $\psi_k[x(n-m)]$ represents the basis function of the orthogonal memory polynomial model and is defined as:

$$\psi_k[x(n-m)] = \sum_{l=1}^{k} (-1)^{l+k} \cdot \frac{(k+l)!}{(l-1)!(l+1)!(k-l)!} \cdot |x(n-m)|^{l-1} \cdot x(n-m) \quad (5.9)$$

The orthogonal memory polynomial model can be re-written in the generic formulation of Equation 5.1 as:

$$y_{OMP}(n) = \boldsymbol{\varphi}_{OMP}(n) \cdot \mathbf{A} \qquad (5.10)$$

where $\mathbf{A}$ is the model coefficients' vector defined as in Equation 5.5. However, in this case, the data vector $\boldsymbol{\varphi}_{OMP}(n)$ is expressed, using the orthogonal memory polynomial basis function defined in Equation 5.9, by:

$$\boldsymbol{\varphi}_{OMP}(n) = \begin{bmatrix} \psi_1[x(n)] \\ \vdots \\ \psi_K[x(n)] \\ \psi_1[x(n-1)] \\ \vdots \\ \psi_K[x(n-1)] \\ \vdots \\ \psi_K[x(n-M)] \end{bmatrix}^T \qquad (5.11)$$

Figure 5.3 presents the block diagram of the orthogonal memory polynomial model. This shows that the model structure is similar to that of the memory polynomial model, and that the only difference lies in the expression of the basis functions adopted. Alternate orthogonal basis functions, such as Zernike polynomials [5, 6] and Gegenbauer polynomials [7], have also been proposed for the modeling and predistortion of power amplifiers and transmitters exhibiting memory effects.

5.4.2   Sparse-Delay Memory Polynomial Model

The memory polynomial model of Equation 5.2 employs polynomial functions that are applied to the current input sample, $x(n)$, as well as all M preceding samples from $x(n-1)$ up to $x(n-M)$. However, all these past terms can be cumbersome, especially when long-term memory effects are being modeled or when the sampling rate of the signal is relatively high and requires a large number of past samples to cover the

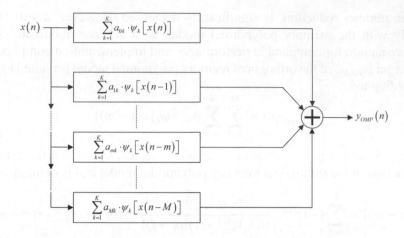

Figure 5.3   Block diagram of the orthogonal memory polynomial model

memory depth of the system being modeled. This motivated the development of the sparse-delay memory polynomial model [8]. In this model, only a specific subset of the preceding samples is taken into account to calculate the output signal. Thus, the output (y_{SDMP}) of the sparse-delay memory polynomial model is given as a function of its input (x) by:

$$y_{SDMP}(n) = \sum_{i=0}^{M_{SD}} \sum_{k=1}^{K} a_{ik} \cdot x(n-m_i) \cdot |x(n-m_i)|^{k-1} \tag{5.12}$$

where a_{ik} and K represents the model's coefficients and nonlinearity order, respectively; and M_{SD} is the number of history terms used to build the sparse-delay memory polynomial model.

According to the description previously, $M_{SD} \leq M$ (where M is the memory depth of the system being modeled) and $m_i \in [0, M]$. The sparse-delay memory polynomial model can be re-written using the generic formulation of Equation 5.1:

$$y_{SDMP}(n) = \boldsymbol{\varphi}_{SDMP}(n) \cdot \mathbf{A} \tag{5.13}$$

where $\mathbf{A}$ is the model coefficients' vector defined as in Equation 5.5. According to Equation 5.12, the data vector $\boldsymbol{\varphi}_{SDMP}(n)$ of the sparse-delay memory polynomial model is:

$$\boldsymbol{\varphi}_{SDMP}(n) = \begin{bmatrix} x(n) \\ \vdots \\ x(n) \cdot |x(n)|^{K-1} \\ x(n-m_1) \\ \vdots \\ x(n-m_1) \cdot |x(n-m_1)|^{K-1} \\ \vdots \\ x(n-m_{M_{SD}}) \cdot |x(n-m_{M_{SD}})|^{K-1} \end{bmatrix}^T \tag{5.14}$$

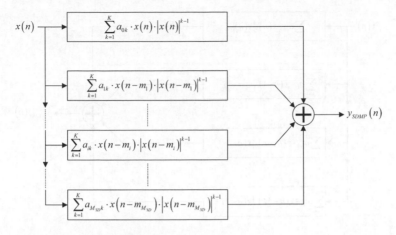

$$x(n)$$

$$\sum_{k=1}^{K} a_{0k} \cdot x(n) \cdot |x(n)|^{k-1}$$

$$\sum_{k=1}^{K} a_{1k} \cdot x(n-m_1) \cdot |x(n-m_1)|^{k-1}$$

$$\sum_{k=1}^{K} a_{ik} \cdot x(n-m_i) \cdot |x(n-m_i)|^{k-1}$$

$$\sum_{k=1}^{K} a_{M_{SD}k} \cdot x(n-m_{M_{SD}}) \cdot |x(n-m_{M_{SD}})|^{k-1}$$

$$y_{SDMP}(n)$$

Figure 5.4   Block diagram of the sparse-delay memory polynomial model

Figure 5.4 shows the block diagram of the sparse-delay memory polynomial model. This model is very similar to the memory polynomial model, except that only a subset of branches is used. The sparse-delay memory polynomial model reduces the complexity of the memory polynomial model by decreasing the number of coefficients present in the model. However, additional complexity is introduced since the sparse-delay memory polynomial model requires the determination of the pertinent delay terms that must be kept in the model and the superfluous terms that can be discarded.

5.4.3   Exponentially Shaped Memory Delay Profile Memory Polynomial Model

The memory polynomial based shaped delay (MPSD) model was presented in [9]. In this model, each branch is assigned a different memory depth to include both thermal and electrical memory effects of the system. Hence, the delay is an exponential function rather than a uniform function in which equal delay steps is incorporated in all the branches. The mathematical representation of the MPSD model is given in Equation 5.15:

$$y_{MPSD}(n) = \sum_{m=0}^{M} \sum_{k=1}^{K} a_{mk} \cdot x(n - \Delta_{m,k}) \cdot |x(n - \Delta_{m,k})|^{k-1} \qquad (5.15)$$

where the delay values are defined as:

$$\Delta_{m,k} = \begin{cases} 0 & m = 0 \\ avg + \Delta_0 e^{-\alpha k} & m \neq 0 \end{cases} \qquad (5.16)$$

In Equation 5.16, Δ_0 is the maximum delay, α is a coefficient describing the exponential decrease of the memory depth value, and avg is the average delay value that is related to the drain-source average channel length. The value the parameter avg of depends on the average power applied to the transistor.

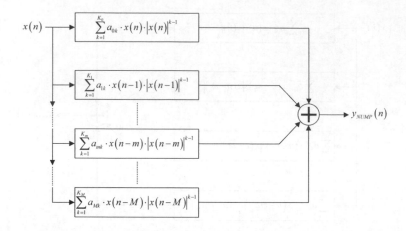

Figure 5.5   Block diagram of the non-uniform memory polynomial model

The advantage of the MPSD model is that there is no need to keep adding branches as in MP model in order to increase the over-all memory depth of the model. In contrast, if a higher delay value is needed in the memory polynomial model with unit delay, extra delay branches are needed, resulting in much greater complexity. The MPSD model extraction requires the calculation of delay coefficients and the model coefficients extraction simultaneously. Experimental validation of this model demonstrated that as the value of M is changed from 0 to 3 the accuracy of the model is substantially enhanced [9].

5.4.4   Non-Uniform Memory Polynomial Model

In the memory polynomial model described by Equation 5.2, all $(M + 1)$ polynomial functions have equal nonlinearity orders. This provides the model with a uniform structure. In the non-uniform memory polynomial (NUMP) model, the nonlinearity orders of the branches are unequal, as illustrated in Figure 5.5. By incorporating independent nonlinearity orders for each branch, Equation 5.2 becomes:

$$y_{NUMP}(n) = \sum_{m=0}^{M} \sum_{k=1}^{K_m} a_{mk} \cdot x(n-m) \cdot |x(n-m)|^{k-1} \tag{5.17}$$

where a_{mk} and M represents the model's coefficients and memory depth, respectively; and, K_m refers to the nonlinearity order of the $(m + 1)th$ branch (associated with the input samples $x(n - m)$) of the non-uniform memory polynomial model.

Taking into account the fact that the nonlinearity order decays for samples associated with deeper memory indices, it is possible to trim down the memory polynomial model by reducing the nonlinearity order of each branch using the following constraint:

$$\begin{cases} K_0 = K \\ \text{for } m_i \text{ and } m_j \in [0, M], \text{ if } m_i \geq m_j \text{ then } K_{m_i} \leq K_{m_j} \end{cases} \tag{5.18}$$

where K and M are the nonlinearity order of the uniform memory polynomial model and its memory depth, respectively.

The generic formulation of Equation 5.1 can also be used for the non-uniform memory polynomial model as:

$$y_{NUMP}(n) = \boldsymbol{\varphi}_{NUMP}(n) \cdot \mathbf{A} \tag{5.19}$$

where $\mathbf{A}$ is the model coefficients' vector defined as in Equation 5.5. According to Equation 5.17, the data vector $\boldsymbol{\varphi}_{NUMP}(n)$ of the non-uniform memory polynomial model is:

$$\boldsymbol{\varphi}_{NUMP}(n) = \begin{bmatrix} x(n) \\ \vdots \\ x(n) \cdot |x(n)|^{K_0-1} \\ x(n-1) \\ \vdots \\ x(n-1) \cdot |x(n-1)|^{K_1-1} \\ \vdots \\ x(n-M) \cdot |x(n-M)|^{K_M-1} \end{bmatrix}^T \tag{5.20}$$

The non-uniform memory polynomial model results in a substantial decrease in the number of coefficients compared to the memory polynomial model [10, 11]. This complexity reduction is more important for wideband applications, where the memory depth of the system is increased.

5.4.5 Unstructured Memory Polynomial Model

The sparse-delay memory polynomial model and the non-uniform memory polynomial model aim both at reducing the total number of coefficients in the memory polynomial model. While in the case of sparse-delay memory polynomial model the selection of the subset of branches is not straightforward, the minimization of the nonlinearity orders used in the non-uniform memory polynomial model can be automated by implementing the criterion of Equation 5.18. The concept of pruning the memory polynomial model can be further extended by simultaneously minimizing the nonlinearity order in each branch as well as the subset of branches to be used. In general, one can minimize the number of coefficients of the memory polynomial model by selecting a subset of coefficients in an unstructured manner using advanced algorithms such as genetic algorithms [12], particle swarm optimization [13], or the adaptive basis function concept [14].

The sparse-delay, the non-uniform, and the unstructured memory polynomial models are considered to be complexity reduced versions of the memory polynomial model. However, the complexity reduction obtained by decreasing the number of coefficients in the model is often accompanied by degradation in the model performance. A trade-off is thus needed in order to reduce the complexity of the model while keeping its performance within a reasonable tolerance range.

$$y_{MP}(n) = a_{01} \cdot x(n) + a_{02} \cdot x(n) \cdot |x(n)| + a_{03} \cdot x(n) \cdot |x(n)|^2 + a_{04} \cdot x(n) \cdot |x(n)|^3 + a_{05} \cdot x(n) \cdot |x(n)|^4$$
$$+ a_{11} \cdot x(n-1) + a_{12} \cdot x(n-1) \cdot |x(n-1)| + a_{13} \cdot x(n-1) \cdot |x(n-1)|^2 + a_{14} \cdot x(n-1) \cdot |x(n-1)|^3 + a_{15} \cdot x(n-1) \cdot |x(n-1)|^4$$
$$+ a_{21} \cdot x(n-2) + a_{22} \cdot x(n-2) \cdot |x(n-2)| + a_{23} \cdot x(n-2) \cdot |x(n-2)|^2 + a_{24} \cdot x(n-2) \cdot |x(n-2)|^3 + a_{25} \cdot x(n-2) \cdot |x(n-2)|^4$$
$$+ a_{31} \cdot x(n-3) + a_{32} \cdot x(n-3) \cdot |x(n-3)| + a_{33} \cdot x(n-3) \cdot |x(n-3)|^2 + a_{34} \cdot x(n-3) \cdot |x(n-3)|^3 + a_{35} \cdot x(n-3) \cdot |x(n-3)|^4$$

$$y_{SDMP}(n) = a_{01} \cdot x(n) + a_{02} \cdot x(n) \cdot |x(n)| + a_{03} \cdot x(n) \cdot |x(n)|^2 + a_{04} \cdot x(n) \cdot |x(n)|^3 + a_{05} \cdot x(n) \cdot |x(n)|^4$$
$$+ a_{11} \cdot x(n-1) + a_{12} \cdot x(n-1) \cdot |x(n-1)| + a_{13} \cdot x(n-1) \cdot |x(n-1)|^2 + a_{14} \cdot x(n-1) \cdot |x(n-1)|^3 + a_{15} \cdot x(n-1) \cdot |x(n-1)|^4$$
$$+ a_{21} \cdot x(n-2) + a_{22} \cdot x(n-2) \cdot |x(n-2)| + a_{23} \cdot x(n-2) \cdot |x(n-2)|^2 + a_{24} \cdot x(n-2) \cdot |x(n-2)|^3 + a_{25} \cdot x(n-2) \cdot |x(n-2)|^4$$
$$+ a_{31} \cdot x(n-3) + a_{32} \cdot x(n-3) \cdot |x(n-3)| + a_{33} \cdot x(n-3) \cdot |x(n-3)|^2 + a_{34} \cdot x(n-3) \cdot |x(n-3)|^3 + a_{35} \cdot x(n-3) \cdot |x(n-3)|^4$$

$$y_{NUMP}(n) = a_{01} \cdot x(n) + a_{02} \cdot x(n) \cdot |x(n)| + a_{03} \cdot x(n) \cdot |x(n)|^2 + a_{04} \cdot x(n) \cdot |x(n)|^3 + a_{05} \cdot x(n) \cdot |x(n)|^4$$
$$+ a_{11} \cdot x(n-1) + a_{12} \cdot x(n-1) \cdot |x(n-1)| + a_{13} \cdot x(n-1) \cdot |x(n-1)|^2 + a_{14} \cdot x(n-1) \cdot |x(n-1)|^3 + a_{15} \cdot x(n-1) \cdot |x(n-1)|^4$$
$$+ a_{21} \cdot x(n-2) + a_{22} \cdot x(n-2) \cdot |x(n-2)| + a_{23} \cdot x(n-2) \cdot |x(n-2)|^2 + a_{24} \cdot x(n-2) \cdot |x(n-2)|^3 + a_{25} \cdot x(n-2) \cdot |x(n-2)|^4$$
$$+ a_{31} \cdot x(n-3) + a_{32} \cdot x(n-3) \cdot |x(n-3)| + a_{33} \cdot x(n-3) \cdot |x(n-3)|^2 + a_{34} \cdot x(n-3) \cdot |x(n-3)|^3 + a_{35} \cdot x(n-3) \cdot |x(n-3)|^4$$

$$y_{UMP}(n) = a_{01} \cdot x(n) + a_{02} \cdot x(n) \cdot |x(n)| + a_{03} \cdot x(n) \cdot |x(n)|^2 + a_{04} \cdot x(n) \cdot |x(n)|^3 + a_{05} \cdot x(n) \cdot |x(n)|^4$$
$$+ a_{11} \cdot x(n-1) + a_{12} \cdot x(n-1) \cdot |x(n-1)| + a_{13} \cdot x(n-1) \cdot |x(n-1)|^2 + a_{14} \cdot x(n-1) \cdot |x(n-1)|^3 + a_{15} \cdot x(n-1) \cdot |x(n-1)|^4$$
$$+ a_{21} \cdot x(n-2) + a_{22} \cdot x(n-2) \cdot |x(n-2)| + a_{23} \cdot x(n-2) \cdot |x(n-2)|^2 + a_{24} \cdot x(n-2) \cdot |x(n-2)|^3 + a_{25} \cdot x(n-2) \cdot |x(n-2)|^4$$
$$+ a_{31} \cdot x(n-3) + a_{32} \cdot x(n-3) \cdot |x(n-3)| + a_{33} \cdot x(n-3) \cdot |x(n-3)|^2 + a_{34} \cdot x(n-3) \cdot |x(n-3)|^3 + a_{35} \cdot x(n-3) \cdot |x(n-3)|^4$$

Figure 5.6   Structure of memory polynomial model variants

Figure 5.6 shows an example that illustrates the difference between the structure of the memory polynomial model and that of its three complexity reduced versions. The memory polynomial model shown in this figure has a memory depth of $M = 3$ and a nonlinearity order of $K = 5$. In the sparse-delay memory polynomial model, only the polynomial branches corresponding to memory depths of $m_i \in \{0, 2, 3\}$ are kept. However, in each of these branches, the nonlinearity order is kept unchanged ($K = 5$). In the non-uniform memory polynomial model, all four branches are kept; however, the nonlinearity order of the branches are customized ($K_0 = 5$, $K_1 = 4$, $K_2 = 2$, $K_3 = 1$). Finally, in the unstructured memory polynomial model, a subset of coefficients is selected.

5.5   Envelope Memory Polynomial Model

The envelope memory polynomial model, which was introduced in [15], employs a basis function different from that of the memory polynomial model, that is, $x(n) \cdot$

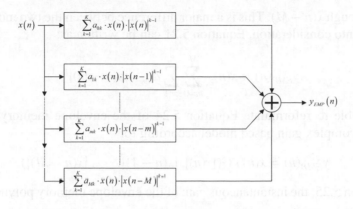

Figure 5.7   Block diagram of the envelope memory polynomial model

$|x(n - m)|^{k-1}$ rather than $x(n - m) \cdot |x(n - m)|^{k-1}$. In the envelope memory polyno-mial model, the output signal (y_{EMP}) is related to the input signal (x) by:

$$y_{EMP}(n) = \sum_{m=0}^{M} \sum_{k=1}^{K} a_{mk} \cdot x(n) \cdot |x(n - m)|^{k-1} \tag{5.21}$$

where a_{mk} are the model's coefficients; and, M and K are the memory depth and non-linearity order, respectively.

The block diagram of the envelope memory polynomial model as described in Equation 5.21 is shown in Figure 5.7.

Similarly to the previous memory polynomial models, the envelope memory poly-nomial model can be reformulated according to the generic equation of Equation 5.1 as follows:

$$y_{EMP}(n) = \varphi_{EMP}(n) \cdot A \tag{5.22}$$

where **A** is the model coefficients' vector defined as in Equation 5.5.

By combining Equations 5.21 and 5.22, the data vector, $\varphi_{EMP}(n)$, of the envelope memory polynomial model can be defined as:

$$\varphi_{EMP}(n) = \begin{bmatrix} x(n) \\ \vdots \\ x(n) \cdot |x(n)|^{K-1} \\ x(n) \\ \vdots \\ x(n) \cdot |x(n-1)|^{K-1} \\ \vdots \\ x(n) \cdot |x(n-M)|^{K-1} \end{bmatrix}^{T} \tag{5.23}$$

Contrary to the memory polynomial model, the envelope memory polynomial model does not require access to the complex values of the previous input samples, that is,

$x(n-1)$ through $x(n-M)$. This is a major difference between the two models. Taking this aspect into consideration, Equation 5.21 can be written as:

$$y_{EMP}(n) = x(n) \cdot \sum_{m=0}^{M} \sum_{k=1}^{K} a_{mk} \cdot |x(n-m)|^{k-1} \tag{5.24}$$

It is possible to reformulate Equation 5.24 of the envelope memory polynomial model as a complex gain based model according to:

$$y_{EMP}(n) = x(n) \cdot G(|x(n)|, |x(n-1)|, \cdots, |x(n-M)|) \tag{5.25}$$

In Equation 5.25, the instantaneous gain of the envelope memory polynomial model is given by:

$$G(|x(n)|, |x(n-1)|, \cdots, |x(n-M)|) = \sum_{m=0}^{M} \sum_{k=1}^{K} a_{mk} \cdot |x(n-m)|^{k-1} \tag{5.26}$$

An alternate representation of the envelope memory polynomial shown in Figure 5.7 is presented in Figure 5.8, where the model is built using a complex gain multiplier structure. This implementation makes the envelope memory polynomial model highly attractive for memory effect compensation using RF digital predistortion systems. In fact, in such systems the designer has access to the RF signal. Accordingly, the instantaneous gain of the envelope memory polynomial model can be built by sensing the amplitude of the input samples using an envelope detector without having to demodulate these RF signals to recover their complex baseband versions.

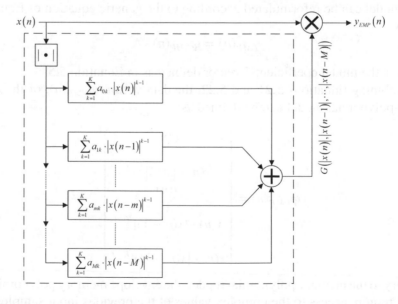

Figure 5.8   Complex gain based representation of the envelope memory polynomial model

The envelope memory polynomial model performance is comparable to that of the memory polynomial model for mildly nonlinear systems. Moreover, the complexity of both models is identical for a given nonlinearity order and memory depth. However, the envelope memory polynomial performance is worse for highly nonlinear systems than that of the memory polynomial model. Although this might impact the use of the envelope memory polynomial as a standalone model, its implementation in hybrid multi-basis function models is highly attractive, as will be discussed in Section 5.7.

5.6   Generalized Memory Polynomial Model

The memory polynomial models presented so far are considered as single basis function models. The generalized memory polynomial model is built by augmenting the memory polynomial model with additional basis functions, which introduces cross-terms that result from combining the instantaneous complex signal with leading and lagging terms [16]. The output (y_{GMP}) of the generalized memory polynomial model is related to its input (x) by:

$$y_{GMP}(n) = \sum_{m=0}^{M_a} \sum_{k=1}^{K_a} a_{mk} \cdot x(n-m) \cdot |x(n-m)|^{k-1}$$

$$+ \sum_{m=0}^{M_b} \sum_{k=2}^{K_b} \sum_{p=1}^{P} b_{mkp} \cdot x(n-m) \cdot |x(n-m-p)|^{k-1}$$

$$+ \sum_{m=0}^{M_c} \sum_{k=2}^{K_c} \sum_{q=1}^{Q} c_{mkq} \cdot x(n-m) \cdot |x(n-m+q)|^{k-1} \tag{5.27}$$

The output (y_{GMP}) of the generalized memory polynomial model is composed of three polynomial functions. The first is applied to time-aligned input signal samples, and has a nonlinearity order and memory depth of K_a and M_a, respectively. The second polynomial function is applied to the complex input signal and lagging values of its envelope. This polynomial function introduces cross-terms between the input signal and its lagging envelope terms up to the Pth order with a nonlinearity order of K_b and a memory depth of M_b. Similarly, cross-terms between the input signal and the leading envelope terms up to the Qth order are introduced through the third polynomial function. The nonlinearity order and memory depth of the leading cross-terms polynomial are K_c and M_c, respectively. In Equation 5.27, a_{mk}, b_{mkp}, and c_{mkq} are the coefficients of the memory polynomial functions applied on the aligned terms, and the lagging and leading cross-terms, respectively.

Typically, the nonlinearity orders and memory depths of the polynomial functions associated with the leading and lagging cross-terms (K_b, M_b, K_c, and M_c) are significantly less than that of the memory polynomial function applied to the time-aligned signal and its envelope (K_a and M_a). Furthermore, the orders of the lagging and leading cross-terms (P and Q, respectively) that need to be taken into account by the model are commonly relatively low.

The block diagram of the generalized memory polynomial model is presented in Figure 5.9. This figure clearly illustrates that the generalized memory polynomial model can be perceived as a memory polynomial model augmented with the combination of $P + Q$ memory polynomial functions, P of which are associated with the lagging cross-terms, and Q of which are associated with the leading cross-terms.

The generalized memory polynomial model can be described by the generic formulation of Equation 5.1 as:

$$y_{GMP}(n) = \boldsymbol{\varphi}_{GMP}(n) \cdot \mathbf{A} \tag{5.28}$$

where $\mathbf{A}$ is the model coefficients' vector and $\boldsymbol{\varphi}_{GMP}(n)$ is the data vector associated with the generalized memory polynomial model. Since the generalized memory polynomial model uses a set of three basis functions, the coefficients as well as the data

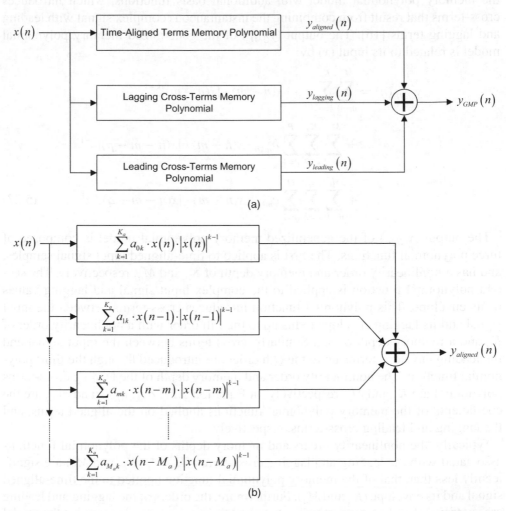

(a)

(b)

Figure 5.9 Block diagram of the generalized memory polynomial model. (a) Generalized memory polynomial model. (b) Time-aligned terms memory polynomial

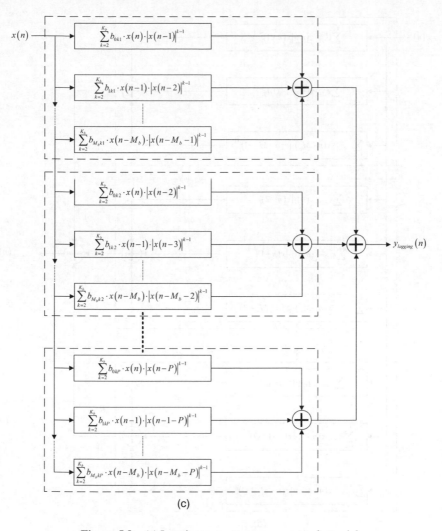

$$\text{(c)}$$

Figure 5.9   (c) Lagging cross-terms memory polynomial

vectors $\mathbf{A}$ and $\boldsymbol{\varphi}_{GMP}(n)$, can be built by concatenating three vectors. $\mathbf{A}$ is defined as:

$$\mathbf{A} = \begin{bmatrix} \mathbf{A}_{aligned} \\ \mathbf{A}_{lagging} \\ \mathbf{A}_{leading} \end{bmatrix} \tag{5.29}$$

with

$$\mathbf{A}_{aligned} = \begin{bmatrix} a_{01} & \cdots & a_{0K_a} & a_{11} & \cdots & a_{1K_a} & \cdots & a_{M_aK_a} \end{bmatrix}^T \tag{5.30}$$

$$\mathbf{A}_{lagging} = \begin{bmatrix} b_{021} & \cdots & b_{02P} & b_{031} & \cdots & b_{03P} & \cdots & b_{M_bK_bP} \end{bmatrix}^T \tag{5.31}$$

$$\mathbf{A}_{leading} = \begin{bmatrix} c_{021} & \cdots & c_{02Q} & c_{031} & \cdots & c_{03Q} & \cdots & c_{M_cK_cQ} \end{bmatrix}^T \tag{5.32}$$

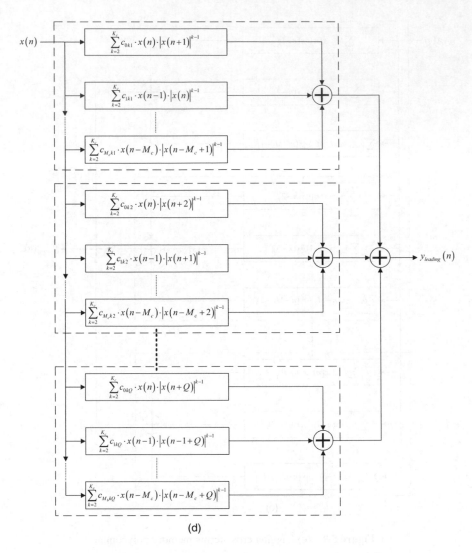

(d)

Figure 5.9   (d) Leading cross-terms memory polynomial

Similarly, by combining Equations 5.27 and 5.28, it is possible to write the data vector $\varphi_{GMP}(n)$ as:

$$\varphi_{GMP}(n) = \begin{bmatrix} \varphi_{aligned}(n) & \varphi_{lagging}(n) & \varphi_{leading}(n) \end{bmatrix} \qquad (5.33)$$

where $\varphi_{aligned}(n)$ represents the data vector constructed using the basis function associated with the signal and its aligned envelope terms, and $\varphi_{lagging}(n)$ and $\varphi_{leading}(n)$ are the data vector made of the signal and its lagging and leading envelope terms, respectively. These data vectors are given by:

$$
\varphi_{aligned}(n) =
\begin{bmatrix}
x(n) \\
\vdots \\
x(n) \cdot |x(n)|^{K_a-1} \\
x(n-1) \\
\vdots \\
x(n-1) \cdot |x(n-1)|^{K_a-1} \\
\vdots \\
x(n-M_a) \cdot |x(n-M_a)|^{K_a-1}
\end{bmatrix}^T
\tag{5.34}
$$

$$
\varphi_{lagging}(n) =
\begin{bmatrix}
x(n) \cdot |x(n-1)| \\
\vdots \\
x(n) \cdot |x(n-P)| \\
x(n) \cdot |x(n-1)|^2 \\
\vdots \\
x(n) \cdot |x(n-P)|^2 \\
\vdots \\
x(n-M_b) \cdot |x(n-M_b-P)|^{K_b-1}
\end{bmatrix}^T
\tag{5.35}
$$

$$
\varphi_{leading}(n) =
\begin{bmatrix}
x(n) \cdot |x(n+1)| \\
\vdots \\
x(n) \cdot |x(n+Q)| \\
x(n) \cdot |x(n+1)|^2 \\
\vdots \\
x(n) \cdot |x(n+Q)|^2 \\
\vdots \\
x(n-M_c) \cdot |x(n-M_c+Q)|^{K_c-1}
\end{bmatrix}^T
\tag{5.36}
$$

The generalized memory polynomial model is attractive when the device under test (DUT) exhibits strong nonlinear memory effects. In such conditions, the generalized memory polynomial outperforms the memory polynomial model; and, the additional complexity associated with the generalized memory polynomial is justified. However, when linear memory effects are present both models lead to comparable performances.

By introducing cross-terms, the generalized memory polynomial model is positioned between the memory polynomial model and the Volterra series, in terms of complexity and performance. It is perceived as a practical alternative to the computationally heavy Volterra series when strong memory effects are present, especially for multi-carrier non-contiguous signals in which off-carriers are present between the on-carriers being used to transmit the signal.

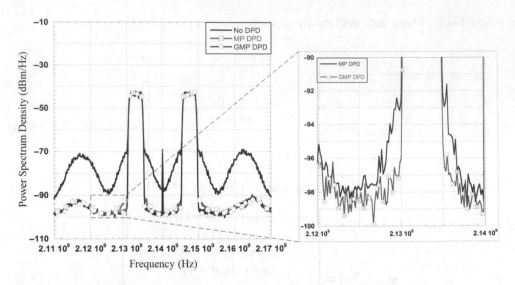

Figure 5.10   Measured spectra at the output of a Doherty amplifier linearized using the memory polynomial and the generalized memory polynomial DPDs

Figure 5.10 presents the spectra measured at the output of a Doherty power amplifier linearized using the memory polynomial and the generalized memory polynomial
model. This figure illustrates the superiority of the generalized memory polynomial
model in reducing the spectral regrowth at the output of the DUT, especially in the
vicinity of the carriers.

5.7   Hybrid Memory Polynomial Model

The hybrid memory polynomial model is conceptually similar to the generalized
memory polynomial model in the sense that both augment the memory polynomial
model through the inclusion of cross-terms. The hybrid memory polynomial model
consists of the parallel arrangement of a memory polynomial model and an envelope
memory polynomial model as illustrated in Figure 5.11 [17]. Thus, the equation
relating the input (x) of the hybrid memory polynomial model to its output (y_{HMP}) is:

$$y_{HMP}(n) = \sum_{m=0}^{M} \sum_{k=1}^{K} a_{mk} \cdot x(n-m) \cdot |x(n-m)|^{k-1}$$

$$+ \sum_{m=1}^{M_e} \sum_{k=2}^{K_e} b_{mk} \cdot x(n) \cdot |x(n-m)|^{k-1} \qquad (5.37)$$

where a_{mk} and b_{mk} are the model's coefficients associated with the memory polynomial and envelope memory polynomial functions; K and M represent the memory

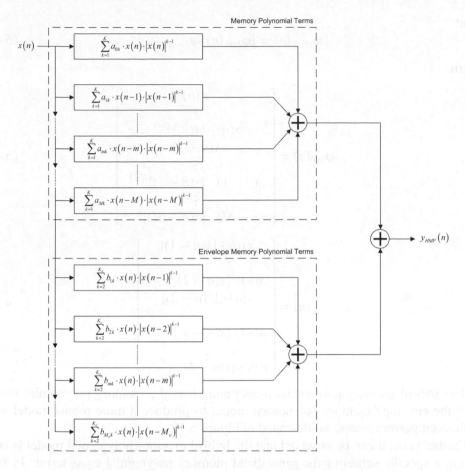

Figure 5.11   Block diagram of the hybrid memory polynomial model

polynomial's function nonlinearity order and the memory depth, respectively; and, equivalently, K_e and M_e represent those of the envelope memory polynomial function.

The hybrid memory polynomial model is built using two basis functions; and thus Equation 5.37 can be rewritten as:

$$y_{HMP}(n) = \boldsymbol{\varphi}_{HMP}(n) \cdot \mathbf{A} \tag{5.38}$$

where the vector coefficients' $\mathbf{A}$ and the data vector $\boldsymbol{\varphi}_{HMP}(n)$ are defined by:

$$\mathbf{A} = \begin{bmatrix} \mathbf{A}_{MP} \\ \mathbf{A}_{EnvMP} \end{bmatrix} \tag{5.39}$$

with

$$\mathbf{A}_{MP} = \begin{bmatrix} a_{01} & \cdots & a_{0K} & a_{11} & \cdots & a_{1K} & \cdots & a_{MK} \end{bmatrix}^T \tag{5.40}$$

$$\mathbf{A}_{EnvMP} = \begin{bmatrix} b_{12} & \cdots & b_{1K_e} & b_{22} & \cdots & b_{2K_e} & \cdots & b_{M_eK_e} \end{bmatrix}^T \tag{5.41}$$

and

$$\boldsymbol{\varphi}_{HMP}(n) = \begin{bmatrix} \boldsymbol{\varphi}_{MP}(n) & \boldsymbol{\varphi}_{EnvMP}(n) \end{bmatrix} \tag{5.42}$$

with

$$\boldsymbol{\varphi}_{MP}(n) = \begin{bmatrix} x(n) \\ \vdots \\ x(n) \cdot |x(n)|^{K-1} \\ x(n-1) \\ \vdots \\ x(n-1) \cdot |x(n-1)|^{K-1} \\ \vdots \\ x(n-M) \cdot |x(n-M)|^{K-1} \end{bmatrix}^{T} \tag{5.43}$$

$$\boldsymbol{\varphi}_{EnvMP}(n) = \begin{bmatrix} x(n) \cdot |x(n-1)| \\ \vdots \\ x(n) \cdot |x(n-1)|^{K_e-1} \\ x(n) \cdot |x(n-2)| \\ \vdots \\ x(n) \cdot |x(n-2)|^{K_e-1} \\ \vdots \\ x(n) \cdot |x(n-M_e)|^{K_e-1} \end{bmatrix}^{T} \tag{5.44}$$

The hybrid memory polynomial model combines the memory polynomial model and the envelope memory polynomial model to produce a more robust model with enhanced performances, as illustrated in Figure 5.12.

Furthermore, it can be observed that the hybrid memory polynomial model is built using a specific subset of the generalized memory polynomial cross-terms. In fact, it only includes cross-terms made of the actual signal sample and its lagging envelope samples. These are expected to be the predominant cross-terms among all those embedded in the generalized memory polynomial model. This explains the fact that the hybrid memory polynomial model often leads to performances comparable to that of the generalized memory polynomial with a reduced number of coefficients.

5.8 Dynamic Deviation Reduction Volterra Model

Various approaches have been proposed to prune the Volterra series model into more compact versions that combine high performance and low complexity, in order to enable its practical use in power amplifier behavioral modeling and predistortion applications. The pruning can be based on the near-diagonality concept [18], the dynamic deviation reduction concept [19], or the radial pruning concept [20]. These versions of the Volterra series include considerably more cross-terms than the generalized and the hybrid memory polynomial models, as they allow for the interaction of several samples having various times indices.

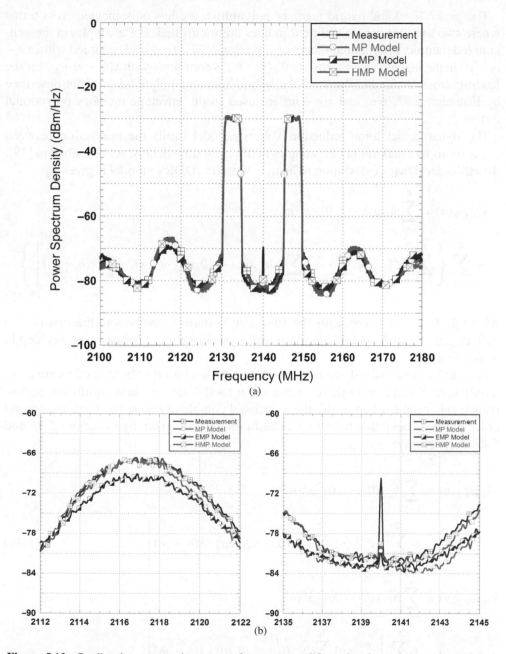

Figure 5.12 Predicted spectra at the output of a power amplifier using the memory polynomial, the envelope memory polynomial, and the hybrid memory polynomial models. (a) Full spectrum and (b) zoomed version

The generalized and hybrid memory polynomial models only include cross-terms where two samples having different indices are multiplied. For example, in the generalized memory polynomial model of Equation 5.27, $x(n - m)$ is coupled with $x(n - m - p)$ in the lagging cross-terms, and $x(n - m)$ is combined with $x(n - m + q)$ for the leading cross-terms. Similarly, in the hybrid memory polynomial model described by Equation 5.37, $x(n)$ and $x(n - m)$ are used in the envelope memory polynomial terms.

The dynamic deviation reduction Volterra model limits the interaction between cross-terms to a maximum of r samples at different time delays. As described in [19], the rth-order dynamic deviation reduction Volterra (DDRV) model is given by:

$$y_{DDRV}(n) = \sum_{k=1}^{K} h_{k,0}(0, \cdots, 0) \cdot x^k(n)$$

$$+ \sum_{k=1}^{K} \left\{ \sum_{r=1}^{k} \left[x^{k-r}(n) \sum_{m_1=1}^{M} \cdots \sum_{m_r=m_{r-1}}^{M} h_{k,r}(0, \cdots, 0, m_1, \cdots, m_r) \cdot \prod_{i=1}^{r} x(n - m_i) \right] \right\}$$

(5.45)

where $h_{k,r}(j_1, \cdots, j_k)$ represents the kth-order Volterra kernel where the first $(k - r)$ indices are 0; and K and M are the model's nonlinearity order and memory depth, respectively.

First and second order dynamic deviation reduction based Volterra models are commonly used. Compared to these, higher order models do not show significant performance enhancement that justify the associated complexity increase. First and second order dynamic deviation reduction Volterra models are given by Equations 5.46 and 5.47, respectively:

$$y_{DDRV,1}(n) = \sum_{k=1}^{K} h_{k,0}(0, \cdots, 0) \cdot x^k(n)$$

$$+ \sum_{k=1}^{K} \left[x^{k-1}(n) \sum_{m=1}^{M} h_{k,1}(0, \cdots, 0, m) \cdot x(n - m) \right]$$

(5.46)

$$y_{DDRV,2}(n) = \sum_{k=1}^{K} h_{k,0}(0, \cdots, 0) \cdot x^k(n)$$

$$+ \sum_{k=1}^{K} \left[x^{k-1}(n) \sum_{m=1}^{M} h_{k,1}(0, \cdots, 0, m) \cdot x(n - m) \right]$$

$$+ \sum_{k=2}^{K} \left[x^{k-2}(n) \sum_{m_1=1}^{M} \sum_{m_2=m_1}^{M} h_{k,2}(0, \cdots, 0, m_1, m_2) \cdot x(n - m_1) \cdot x(n - m_2) \right]$$

(5.47)

5.9 Comparison and Discussion

In this chapter, widely used memory polynomial based single-box behavioral models and digital predistorter structures are described. The use of these structures as part of a multi-box model is discussed in Chapter 6.

Table 5.1 presents the parameters and total number of coefficients of these models. As reported in this table, the memory polynomial model requires the knowledge of the model dimension and the selection of two parameters (nonlinearity order and

Table 5.1 Comparison between memory polynomial models' complexity

Model	Equations	Parameters	Total number of coefficients
Memory Polynomial/ Orthogonal Memory Polynomial	5.2 5.8	K: nonlinearity order M: memory depth	$K \times (M + 1)$
Sparse-Delay Memory Polynomial	5.12	K: nonlinearity order m_i for $i \in [1, M_{SD}]$: sparse delay values	$K \times M_{SD}$
Non-Uniform Memory Polynomial	5.17	M: memory depth K_m for $m \in [0, M]$: nonlinearity order of the mth branch	$\displaystyle\sum_{m=0}^{M} K_m$
Envelope Memory Polynomial	5.21	K: nonlinearity order M: memory depth	$K \times (M + 1)$
Generalized Memory Polynomial	5.27	(K_a, M_a), (K_b, M_b), and (K_c, M_c): nonlinearity order and memory depth of the aligned terms, lagging and leading cross-terms polynomials, respectively P and Q: order of the lagging and leading cross-terms, respectively	$K_a \times (M_a + 1)$ $+ P \times (K_b - 1) \times (M_b + 1)$ $+ Q \times (K_c - 1) \times (M_c + 1)$
Hybrid Memory Polynomial	5.37	(K, M) and (K_e, M_e): nonlinearity order and memory depth of the memory polynomial and envelope memory polynomial sub-function, respectively	$K \times (M + 1) + (K_e - 1) \times M_e$
First Order Dynamic Deviation Reduction Volterra	5.46	K: nonlinearity order M: memory depth	$K \times (M + 1)$
Second Order Dynamic Deviation Reduction Volterra	5.47	K: nonlinearity order M: memory depth	$\left(K + \frac{(K-1) \times M}{2}\right) \times (M + 1)$

memory depth) before identifying the model coefficients. The number of parameters increases up to eight in the case of the generalized memory polynomial model. Accurate determination of these parameters to avoid an oversized model or poor performance is an important but not a straightforward task. In fact, a higher number of coefficients in the model imply that more resources needed for the coefficients identification, as well as the model implementation.

Information theory based techniques have been applied to determine the parameters of the memory polynomial model [21, 22]. Empirical approaches, in which the parameters (e.g., nonlinearity order and memory depth) are increased or swept until satisfactory performances are obtained, can be used to estimate the parameters of the model [23]. However, these become unpractical for models with cross-terms.

Prior knowledge of the DUT or simplifying assumptions can reduce the number of parameters to be estimated. For example, in the hybrid memory polynomial, a reasonable assumption would be to force both memory depths to be equal. Similarly, in the case of the generalized memory polynomial, the memory depth and nonlinearity order of the memory polynomials associated with the leading and lagging cross-terms can be made equal. Such assumptions may result in slightly oversized models but are necessary for the selection of the model parameters.

Each of the memory polynomial and envelope memory polynomial models uses a single basis function. This leads to satisfactory performance when weakly nonlinear memory effects are present in the DUT. As the contribution of nonlinear memory

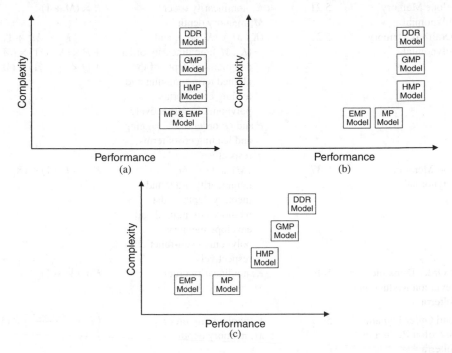

Figure 5.13 Comparison between memory polynomial based models. (a) Weakly nonlinear memory effects, (b) mildly nonlinear memory effects, and (c) strongly nonlinear memory effects

effects gets stronger, the performances of these single basis function models tend to degrade. This justifies the use of more comprehensive models that include cross-terms and thus involve a higher number of coefficients.

Figure 5.13 compares the performance and complexity of the memory polynomial models discussed in this chapter depending on the types of memory effects present in the system being modeled or linearized. A best model that suits any power amplifier cannot be claimed; rather, for each power amplifier, there is the best model that leads to satisfactory accuracy with the lowest possible number of parameters.

The classification provided in Figure 5.13 can be used as a guideline for the selection of the appropriate memory polynomial based behavioral model and digital predistorter. Metrics such as memory effect intensity can be used to determine the strength of the system's memory effects and steer the choice.

References

[1] J. Kim and K. Konstantinou, Digital predistortion of wideband signals based on power amplifier model with memory, *Electronics Letters*, **37**, 23, 1417–1418, 2001.

[2] S. Saied-Bouajina, O. Hammi, M. Jaidane-Saidane, and F. M. Ghannouchi, Experimental approach for robust identification of radiofrequency power amplifier behavioural models using polynomial structures, *IET Microwaves, Antennas & Propagation*, **4**, 11, 1418–1428, 2010.

[3] R. Raich, Q. Hua, and G. T. Zhou, Orthogonal polynomials for power amplifier modeling and predistorter design, *IEEE Transactions on Vehicular Technology*, **53**, 5, 1468–1479, 2004.

[4] O. Hammi, S. Boumaiza, and F. M. Ghannouchi, On the robustness of digital predistortion function synthesis and average power tracking for highly nonlinear power amplifiers, *IEEE Transactions on Microwave Theory and Techniques*, **55**, 6, 1382–1389, 2007.

[5] C. Jebali, N. Boulejfen, M. Rawat, A. Gharsallah, and F. M. Ghannouchi, Modeling of wideband radio frequency power amplifiers using Zernike polynomials, *International Journal of RF and Microwave Computer-Aided Engineering*, **22**, 3, 289–296, 2012.

[6] L. Aladren, P. Garcia-Ducar, P.L. Carro, J. de Mingo, and C. Sanchez-Perez, High power amplifier linearization using Zernike polynomials in a LTE transmission. Proceedings IEEE Vehicular Technology Conference (VTC-Fall), Quebec City, Canada, September 2012, pp. 1–5, 2012.

[7] A. Harguem, N. Boulejfen, F. M. Ghannouchi, and A. Gharsallah, Robust behavioral modeling of dynamic nonlinearities using Gegenbauer polynomials with application to RF power amplifiers, *International Journal of RF and Microwave Computer-Aided Engineering*, **24**, 2, 268–279, 2014.

[8] K. Hyunchul and J. S. Kenney, Behavioral modeling of nonlinear RF power amplifiers considering memory effects, *IEEE Transactions on Microwave Theory and Techniques*, **51**, 12, 2495–2504, 2003.

[9] A. H. Yuzer, S. A. Bassam, F. M. Ghannouchi, and S. Demir, Memory polynomial with shaped memory delay profile and modeling the thermal memory effect in power amplifiers. Proceedings of the IEEE International Conference on Electronics, Circuits, and Systems (ICECS), Abu Dhabi, UAE, December 2013, pp. 1–4, 2013.

[10] O. Hammi, A. M. Kedir, and F. M. Ghannouchi, Nonuniform memory polynomial behavioral model for wireless transmitters and power amplifiers. Proceedings of the 2012 IEEE Asia Pacific Microwave Conference (APMC), Kaohsiung, Taiwan, December 2012, pp. 836–838, 2012.

[11] N. Messaoudi, M. C. Fares, S. Boumaiza, and J. Wood, Complexity reduced odd-order memory polynomial pre-distorter for 400-watt multi-carrier Doherty amplifier linearization. Digest 2008 IEEE MTT-S International Microwave Symposium (IMS), Atlanta, GA, June 2008, pp. 419–422, 2008.

[12] R. Mondal, T. Ristaniemi, and M. Doula, Genetic algorithm optimized memory polynomial digital pre-distorter for RF power amplifiers. Proceedings of the 2013 IEEE International Conference on Wireless Communications & Signal Processing (WCSP), Hangzhou, China, October 2013, pp. 1–5, 2013.

[13] A. H. Abdelhafiz, O. Hammi, A. Zerguine, A. T. Al-Awami, and F. M. Ghannouchi, A PSO based memory polynomial predistorter with embedded dimension estimation, *IEEE Transactions on Broadcasting*, **59**, 4, 665–673, 2013.

[14] Y. Xin and J. Hong, Digital predistortion using adaptive basis functions, *IEEE Transactions on Circuits and Systems I: Regular Papers*, **60**, 12, 3317–3327, 2013.

[15] O. Hammi, F. M. Ghannouchi, and B. Vassilakis, A compact envelope-memory polynomial for RF transmitters modeling with application to baseband and RF-digital predistortion, *IEEE Microwave and Wireless Components Letters*, **18**, 5, 359–361, 2008.

[16] D. R. Morgan, Z. Ma, J. Kim, M. G. Zierdt, and J. Pastalan, A generalized memory polynomial model for digital predistortion of RF power amplifiers, *IEEE Transactions on Signal Processing*, **54**, 10, 3852–3860, 2006.

[17] O. Hammi, M. Younes, and F. M. Ghannouchi, Metrics and methods for benchmarking of RF transmitter behavioral models with application to the development of a hybrid memory polynomial model, *IEEE Transactions on Broadcasting*, **56**, 3, 350–357, 2010.

[18] A. Zhu and T. J. Brazil, Behavioral modeling of RF power amplifiers based on pruned Volterra series, *IEEE Microwave and Wireless Components Letters*, **14**, 12, 563–565, 2004.

[19] A. Zhu, J. C. Pedro, and T. J. Brazil, Dynamic deviation reduction-based Volterra behavioral modeling of RF power amplifiers, *IEEE Transactions on Microwave Theory and Techniques*, **54**, 12, 4323–4332, 2006.

[20] C. Crespo-Cadenas, J. Reina-Tosina, M. J. Madero-Ayora, and J. Munoz-Cruzado, A new approach to pruning Volterra models for power amplifiers, *IEEE Transactions on Signal Processing*, **58**, 4, 2113–2120, 2010.

[21] M. V. Amiri, S. A. Bassam, M. Helaoui, and F. M. Ghannouchi, New order selection technique using information criteria applied to SISO and MIMO systems predistortion, *International Journal of Microwave and Wireless Technologies*. **5**, 2, 123–131, 2013.

[22] J. Wood, M. Lefevre, D. Runton, J. C. Nanan, B. H. Noori, and P. H. Aaen, Envelope-domain time series (ET) behavioral model of a Doherty RF power amplifier for system design, *IEEE Transactions on Microwave Theory and Techniques*, **54**, 8, 3163–3172, 2006.

[23] O. Hammi, M. Younes, A. Kwan, M. Smith, and F. M. Ghannouchi, Performance-driven dimension estimation of memory polynomial behavioral models for wireless transmitters and power amplifiers, *Progress In Electromagnetics Research C*, **12**, 173–189, 2010.

6

Box-Oriented Models

6.1 Introduction

Various behavioral modeling techniques have been introduced and detailed in previous chapters. Chapter 5 explains polynomial based modeling and compensation of non-linear distortions introduced by PAs with reasonably large memory effects. It has been shown that these polynomial models result in high performance. However, implementation of these models results in hardware complexity as these models yield a large number of coefficients. Hence, low complexity is essential for the design of behavioral models and their hardware implementation.

In this chapter, box-/block-oriented models that model the dynamic nonlinear effects of the PA are explained. These models aim to reduce the complexity and enhance the numerical stability of the system. The box-oriented models tend to reduce the number of coefficients required to compute the model and improve the matrix conditioning and dispersion coefficient.

6.2 Hammerstein and Wiener Models

Earlier behavioral models, such as the Saleh model, model the static nonlinear behavior of PAs and do not consider the important memory effects that have gained enormous importance, due to the increasing bandwidth. In this section, two simple box-oriented models for modeling the nonlinear behavior of PAs are described. These models – the Wiener model and the Hammerstein model – represent a class of two-block models where one block accounts for the static nonlinear behavior of the PA, while the other deals with the linear memory effects in the system.

Behavioral Modeling and Predistortion of Wideband Wireless Transmitters, First Edition.
Fadhel M. Ghannouchi, Oualid Hammi and Mohamed Helaoui.
© 2015 John Wiley & Sons, Ltd. Published 2015 by John Wiley & Sons, Ltd.

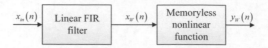

Figure 6.1 Block diagram of the Wiener model

6.2.1 Wiener Model

The Wiener model [1–3] is a concatenation of a linear finite impulse response (FIR) filter and a memoryless nonlinear function, such as the one implemented by a look-up table (LUT), as shown in Figure 6.1. For an input $x_{in}(n)$, the output of the model, $y_W(n)$, is given by [1]:

$$y_W(n) = G_W[|x_W(n)|] \cdot x_W(n) \qquad (6.1)$$

where $G_W[|x_W(n)|] = G_{Wi} + jG_{Wq}$ is the complex memoryless instantaneous gain function of the Wiener model implemented in the LUT model; and, $x_W(n)$ is the output of the FIR filter:

$$x_W(n) = \sum_{m=0}^{M} a_m \cdot x_{in}(n-m) \qquad (6.2)$$

where $x_{in}(n)$ is the input to the system, m indicates the memory of the filter and a_m are the coefficients of the filter, and M is the memory depth.

The construction and identification of the Wiener model requires the removal of dispersion in the transmitter due to the memory effects through extraction of static nonlinearity. This is achieved by fitting the AM/AM (amplitude modulation to amplitude modulation) and AM/PM (amplitude modulation to phase modulation) characteristics of the device under test using a memoryless model by employing one of the identification techniques described in Chapter 8. In this model, the memoryless AM/AM and AM/PM characteristics are often represented using look-up tables.

As mentioned in previous chapters, the PA exhibits memory effects. In order to identify these memory effects, the knowledge of the input signal, that is, $x_{in}(n)$ and $x_W(n)$, is required. However, due to the lack of knowledge of $x_W(n)$ during transmitter characterization, $y_W(n)$ can be used to initialize the time of $x_{in}(n)$. The identification and modeling of these memory effects is also important in behavioral modeling. This is achieved by using an FIR filter. Equations 6.3–6.5 in the form of Equation 5.1 shows the matrix representation of Equation 6.2, which can be written as:

$$\mathbf{X}_W = \mathbf{X}_{in} \cdot \mathbf{A} \qquad (6.3)$$

$$\mathbf{A} = \begin{bmatrix} a_0 & \cdots & a_M \end{bmatrix}^T \qquad (6.4)$$

$$\mathbf{X}_{in} = \begin{bmatrix} x(n) \\ \vdots \\ x(n-M) \end{bmatrix}^T \qquad (6.5)$$

The weights of the filter, $\mathbf{A}$, can be determined by the recursive least squares method [3, 4]. The aim of the method is the reduction of the error between $x(n)$ and $x'(n)$, which is defined as:

$$e(n) = x'(n) - x(n) \qquad (6.6)$$

where $x'(n)$ is the value of the FIR filter at a particular instance.

6.2.2 Hammerstein Model

The Hammerstein model [1, 3] is a combination of a nonlinear memoryless function (such as the one implemented by a LUT) and a linear filter, respectively. Mathematically, such a model takes the following form [1]:

$$y_H(n) = \sum_{m=0}^{M} a_m \cdot x_H(n-m) \qquad (6.7)$$

where

$$x_H(n) = G_H[|x_{in}(n)|] \cdot x_{in}(n) \qquad (6.8)$$

where $x_{in}(n)$ is the input to the system and $y_H(n)$ is the estimated output; $x_H(n)$, a_m, and $G_H[|x_{in}(n)|]$ are the output of the first box (LUT model), the coefficients of the FIR filter and the complex memoryless instantaneous gain of the LUT model, respectively; and, M is the memory depth of the filter.

The identification procedure of the model parameters is similar to that of the Wiener model (the LUT model is identified first, and then the filter coefficients). The model structure is shown in Figure 6.2. The dynamic exponentially weighted moving average method has been employed in the literature [3], to de-embed the static nonlinearity of the PA where the weight factor β is given by:

$$\beta = \frac{\beta_0}{\left\{ \lambda + (1-\lambda) \dfrac{|x_{in}(n)|_{max} - |x_{in}(n)|}{|x_{in}(n)|_{max} - |x_{in}(n)|_{min}} \right\}^p} \qquad (6.9)$$

here β_0 is a constant weight factor and lies between 0 and 1; while, p controls the variation speed of the weighting factor with respect to the input; and λ is the adjustment factor. As with the Wiener model, the coefficients of the model can be extracted using the recursive least squares method.

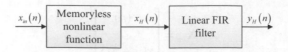

Figure 6.2   Block diagram of the Hammerstein model

6.3   Augmented Hammerstein and Weiner Models

It has been widely shown in the literature that the transmitter exhibits memory effects caused by electrical and thermal dispersion effects, which cannot be ignored in real systems [3]. The various causes of the electrical memory effects have been discussed in [3]. A behavioral model should, therefore, include both linear and nonlinear memory effects for proper modeling and compensation of the dynamic PA nonlinearity.

6.3.1   Augmented Wiener Model

The augmented Wiener model is an extended version of the Wiener model (which uses linear filters) that aims to add a nonlinear memory effect component into the system. In this architecture, a parallel filter branch is added to the linear FIR filter, in which the input is multiplied by its magnitude to form a weak nonlinear filter. This model takes into account the memory effects more appropriately and results in better accuracy than the conventional Wiener model. The block diagram of this extended model is shown in Figure 6.3.

The new output, $x_{AW}(n)$, of the parallel filters can be described by:

$$x_{AW}(n) = \sum_{m_1=0}^{M_1} h_{m_1} \cdot x_{in}(n - m_1)$$

$$+ \sum_{m_2=0}^{M_2} k_{m_2} \cdot x_{in}(n - m_2) \cdot |x_{in}(n - m_2)| \qquad (6.10)$$

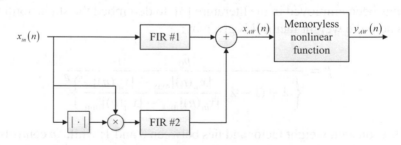

Figure 6.3   Augmented Wiener model

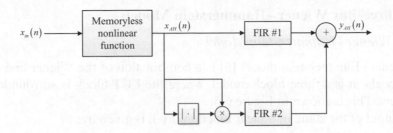

Figure 6.4   Augmented Hammerstein model

where M_1 and M_2 are the memory depth of the first and second filter, respectively; and, h_{m_1} and k_{m_2} are the filter responses.

When Equation 6.10 is compared to Equation 6.2, the effect of the additional FIR filter can be seen. As mentioned earlier, additional memory has been added to the system in order to capture the nonlinear dynamic behavior of the PA more accurately:

$$y_{AW}(n) = G_{AW}(|x_{AW}(n)|) \cdot x_{AW}(n) \tag{6.11}$$

y_{AW} is the output of the model and G_{AW} is the gain of the static nonlinear function modeled by the LUT. The recursive least squares method can also be used in the augmented Wiener model to extract the model parameters. Experimental results for the augmented Wiener model have been presented in [3], where it was shown that the augmented model can predict the memory effects better than the conventional model.

6.3.2   Augmented Hammerstein Model

The augmented Hammerstein model [5] is similar to the augmented Wiener model, but with two parallel branches of the LUT and FIR filters, as shown in Figure 6.4. The output of the LUT box, $x_{AH}(n)$, and the total model output, $y_{AH}(n)$, is:

$$x_{AH}(n) = G_{AH}(|x_{in}(n)|) \cdot x_{in}(n) \tag{6.12}$$

$$y_{AH}(n) = \sum_{m_1=0}^{M_1} h_{m_1} \cdot x_{AH}(n - m_1)$$

$$+ \sum_{m_2=0}^{M_2} k_{m_2} \cdot x_{AH}(n - m_2) \cdot |x_{AH}(n - m_2)| \tag{6.13}$$

y_{AH} is the output of the model and G_{AH} is the gain of the static nonlinear function modeled by the LUT; and, where M_1 and M_2 are the memory depth of the first and second filters, respectively. h_{m_1} and k_{m_2} are the filter responses.

6.4 Three-Box Wiener–Hammerstein Models

6.4.1 Wiener–Hammerstein Model

The Wiener–Hammerstein model [6] is a combination of the Wiener and Hammerstein models. It is a three-block model, where the LUT block is surrounded by two FIR filters. This is shown in Figure 6.5.

The output of the static nonlinear function, $F(\cdot)$, is given by:

$$u_{WH}(n) = F[x_{WH}(n)] = \sum_{i=1}^{N} b_i \cdot x_{WH}(n).|x_{WH}(n)|^{i-1}. \tag{6.14}$$

Here, N and b_i are the nonlinearity order and the coefficients of the nonlinear function $F(\cdot)$, respectively. The outputs of the two FIR filters, that is, $x_{WH}(n)$ and $y_{WH}(n)$, are given by:

$$x_{WH}(n) = \sum_{m_1=0}^{M_1} h_{m_1} \cdot x_{in}(n - m_1) \tag{6.15}$$

and

$$y_{WH}(n) = \sum_{m_2=0}^{M_2} k_{m_2} \cdot u_{WH}(n - m_2) \tag{6.16}$$

M_1 and M_2 are the memory depths of the two FIR filters respectively; while, h_{m_1} and k_{m_2} are the filter responses.

6.4.2 Hammerstein–Wiener Model

The Hammerstein–Wiener model [6] is a combination of the Hammerstein model and the Wiener model. It is also a three-block model, but the FIR filter block is surrounded by two static nonlinear functions, namely $F(\cdot)$ and $G(\cdot)$, as shown in Figure 6.6.

The output of the FIR filter block is given by:

$$u_{HW}(n) = \sum_{m=0}^{M} h_m \cdot x_{HW}(n - m) \tag{6.17}$$

where M and h_m are the memory depth and coefficients of the filter, respectively. The outputs of the two nonlinear functions are given by:

$$x_{HW}(n) = F[x_{in}(n)] = \sum_{i=1}^{N_1} a_i \cdot x_{in}(n) \cdot |x_{in}(n)|^{i-1}, \tag{6.18}$$

$$y_{HW}(n) = G[u_{HW}(n)] = \sum_{j=1}^{N_2} b_j \cdot u_{HW}(n) \cdot |u_{HW}(n)|^{j-1} \tag{6.19}$$

here, N_1 and N_2 are the nonlinearity orders of the nonlinear functions $F(\cdot)$ and $G(\cdot)$, respectively; and a_i and b_j are the coefficients of the nonlinear functions $F(\cdot)$ and

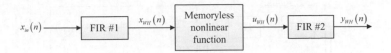

Figure 6.5   Wiener–Hammerstein model

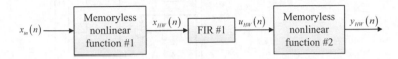

Figure 6.6   Hammerstein–Wiener model

$G(\cdot)$, respectively. In order to maintain uniqueness in the estimation procedure, the first coefficients of the nonlinear function blocks should be kept fixed [6, 7]. This non-uniqueness in the solution comes from the fact that the gain of the system can be arbitrarily divided among the blocks.

Implementation of the model can be done by using the mesh adaptive direct search (MADS) algorithm [6] or the Nelder–Mead algorithm [8]. An analysis of the Hammerstein–Wiener model implementation has been provided in [6], where the authors applied the model by adopting a two-step approach using the MADS algorithm for faster implementation of the model.

The Hammerstein–Wiener model is first divided into a Hammerstein model (the first two blocks) and a static nonlinear function (the third block); and, the parameters of the static nonlinear function are initialized and used as the initial guess for the Hammerstein model. The current output of the Hammerstein model then serves to update the output nonlinearity for the next iteration. Linear least squares methods are then used to minimize the output error. Finally, the MADS algorithm is used to reduce the output error, by using the parameters of the Hammerstein model.

6.4.3   Feedforward Hammerstein Model

The feedforward Hammerstein model [9], as demonstrated in Figure 6.7, adds robustness and enhances the performance of the model by adding more filters to the architecture, which consists of a signal cancelation (SC) loop and a distortion injection (DI) loop. The SC loop is a conventional Hammerstein model for modeling the linear memory effects. However, with the addition of the DI loop, the extended model now accounts for the nonlinear dynamic effects of the PA and is thus more capable of modeling a PA than conventional architectures.

The distortion cancelation loop is composed of a parallel combination of FIR filters. The total model output is given by:

$$y_{FFH}(n) = y_{SC}(n) + y_{DI}(n) \tag{6.20}$$

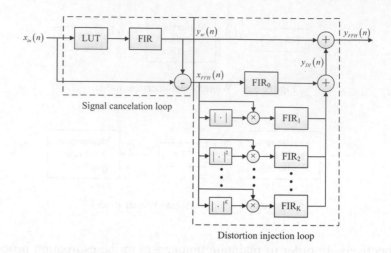

Figure 6.7 Feedforward Hammerstein model

where $y_{FFH}(n)$ is the total output of the model; and, $y_{SC}(n)$ and $y_{DI}(n)$ are the outputs of the SC loop and the DI loop, respectively.

The output of the LUT block, with a gain G_{FH}, is given by:

$$x_{LUT}(n) = G_{FH}(|x_{in}(n)|) \cdot x_{in}(n). \tag{6.21}$$

The outputs of the SC loop and DI loop are given by:

$$y_{SC}(n) = \sum_{m_1=0}^{M_1} a_{m_1} \cdot x_{LUT}(n - m_1), \tag{6.22}$$

$$y_{DI}(n) = \sum_{m_2=0}^{M_2} \sum_{k=0}^{K} b_{m_2 k} \cdot x_{FFH}(n - m_2) \cdot |x_{FFH}(n - m_2)|^k \tag{6.23}$$

where a_{m_1} and $b_{m_2 k}$ represent the coefficients of the FIR filters, M_1 and M_2 represent the memory depths of the blocks, and k is the nonlinearity index of the DI block. Equations 6.21–6.23 can be combined and written in the matrix/vector form given in Equation 5.1.

The coefficients of the model can be obtained by using the least squares approach. Detailed implementation of the architecture, identification process, and experimental results are provided in [9]. It is shown that the feedforward Hammerstein model performs better than the memory polynomial (MP), Hammerstein, and augmented Hammerstein models. Its complexity is decreased compared to the MP model, but is higher than the Hammerstein and augmented Hammerstein models.

6.5   Two-Box Polynomial Models

6.5.1   Models' Descriptions

A two-block model for the modeling and digital predistortion (DPD) of PAs has been proposed [10], for the purpose of reducing the complexity of the system. One of the blocks is a MP based model, while the other is a LUT model. Different arrangements of the blocks lead to the following twin-nonlinear two-box (TNTB) models:

- *Forward-cascaded polynomial model*: In the forward twin-nonlinear two-box model (FTNTB), the LUT is placed before the MP function, as shown in Figure 6.8a.
- *Backward-cascaded polynomial model*: In the backward twin-nonlinear two-box model (BTNTB), the LUT is placed after the MP function, as shown in Figure 6.8b.
- *Parallel-cascaded polynomial model*: In the parallel twin-nonlinear two-box model (PTNTB), the LUT and MP functions are placed parallel to each other, as shown in Figure 6.8c.

The MP model is one type of polynomial that can be used to mimic the dynamic nonlinear behavior of a PA: other polynomial models, such as the Volterra series, can also be used instead of the MP block. Similarly, the LUT is used to model the memoryless nonlinear behavior of the PA; however, other memoryless polynomial functions can be used instead.

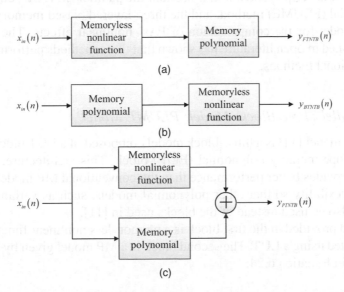

Figure 6.8   Twin nonlinear twin box models. (a) Forward TNTB model, (b) reverse TNTB model, and (c) parallel TNTB model

6.5.2   Identification Procedure

As mentioned earlier, TNTB models have an advantage over conventional polynomial models, as they reduce the complexity of the system. The reason for this is that the highly nonlinear memoryless behavior and the mildly nonlinear memory effects are modeled separately, thereby decreasing the number of coefficients. The identification procedure for the coefficients is composed of two steps:

1. The coefficients for the nonlinear memoryless behavior (modeled by an LUT) of the device under test are obtained.
2. The coefficients for the dynamic nonlinear behavior (modeled by a MP) are then obtained.

The proposed model has been tested with a 300-watt Doherty PA and a four-carrier WCDMA (Wideband Code Division Multiple Access) 1001 test signal [10]. It was shown that the complexity of the system was reduced by 50%, while improving the normalized mean square error (NMSE) by 2 dB.

6.6   Three-Box Polynomial Models

With increasing developments in technology, many models have been proposed to improve the performance of the DPD models. Two such techniques that also reduce the complexity of the system, which is also an important attribute of a behavioral modeling system, are presented in this section: the parallel LUT, MP, envelope memory polynomial (PLUME) method, and the three-layered biased memory polynomial (TLBMP) model for the compensation of PAs' nonlinear effects. The comparative results published in open literature has shown that these methods perform much better than conventional methods.

6.6.1   Parallel Three-Blocks Model: PLUME Model

The PLUME model [11] is a three-block model composed of a LUT block, MP block, and an envelope memory polynomial (EMP) block. This architecture, as shown in Figure 6.9, provides better performance than the conventional MP model. It also provides more flexibility so that other polynomial models, such as variants of the MP model, can also be used instead of the blocks used in [11].

The method provided in the first block is a memoryless nonlinear function, that can be implemented using a LUT. The second block is a MP model given by Equation 5.2 as reiterated in Equation 6.24:

$$y_{MP}(n) = \sum_{m=0}^{M} \sum_{k=1}^{K} a_{mk} \cdot x_{in}(n-m) \cdot |x_{in}(n-m)|^{k-1} \qquad (6.24)$$

M and K indicate the memory depth and the nonlinearity order of the model respectively. The third block is an envelope MP block given by Equation 5.21, which is

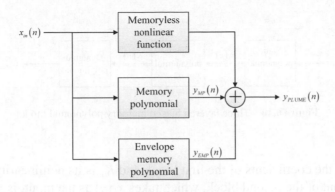

Figure 6.9 PLUME model

restated as in Equation 6.25:

$$y_{EMP}(n) = \sum_{m=0}^{M} \sum_{k=1}^{K} b_{mk} \cdot x_{in}(n) \cdot |x_{in}(n-m)|^{k-1}. \tag{6.25}$$

The modeling identification of y_{MP} and y_{EMP} can be considered as a linear identification problem:

$$Y_{PLUME} = \begin{bmatrix} \Theta_{MP} & \Theta_{EMP} \end{bmatrix} \cdot A = \Theta_{PLUME} \cdot A \tag{6.26}$$

where A is the vector of coefficients, which can be obtained by using the linear least squares method and can be given by:

$$A = (\Theta_{PLUME}^{H} \Theta_{PLUME})^{-1} \cdot \Theta_{PLUME}^{H} \cdot Y_{PLUME} \tag{6.27}$$

Experimental validation was provided in [11] for a Doherty PA with a peak power of 300 W operating in the frequency range of 2110–2170 MHz. The test signal was a 20 MHz WCDMA 1001 signal. It was shown that the PLUME model had a better performance than the conventional MP model (Section 5.3) and the PTNTB model discussed in Section 6.5 and a similar performance to that of the generalized MP model (Section 5.6). However, the major advantage of the PLUME model is the reduction in the complexity of the system in terms of number of coefficients, which makes it highly suitable for implementation in a field-programmable gate array (FPGA).

6.6.2   Three Layered Biased Memory Polynomial Model

The TLBMP model [12] is a cascade of two static polynomial models and a dynamic polynomial model, as illustrated in Figure 6.10.

For input signal $x_{in}(n)$, the output of the first block is given by:

$$x_1(n) = \sum_{k_1=1}^{K_1} a_{k_1} \cdot x_{in}(n) \cdot |x_{in}(n)|^{k_1-1} \tag{6.28}$$

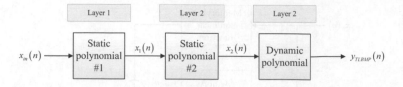

Figure 6.10   Three layered biased memory polynomial model

where a_{k_1} are the coefficients of the first block and K_1 is its nonlinearity order. Similarly, the output of the second block, which takes $x_1(n)$ as the input, is given by:

$$x_2(n) = \sum_{k_2=1}^{K_2} b_{k_2} \cdot x_1(n) \cdot |x_1(n)|^{k_2-1} + bias \qquad (6.29)$$

where b_{k_2} are the coefficients of the second block and K_2 is its nonlinearity order. The final output of the model is given by:

$$y_{TLBMP}(n) = \sum_{k_3=1}^{K_3} \sum_{m=0}^{M} c_{k_3m} \cdot x_2(n-m) \cdot |x_2(n-m)|^{k_3-1} \qquad (6.30)$$

where c_{k_3m} are the coefficients of the third block and K_3 and M are its nonlinearity order and memory depth, respectively.

In matrix notation, these equations (6.28–6.30) can be written in a similar way to Equation 5.1 as:

$$\mathbf{X_1} = \mathbf{X_{in}} \cdot \mathbf{A} \qquad (6.31)$$

$$\mathbf{X_2} = \mathbf{X_1} \cdot \mathbf{B} \qquad (6.32)$$

and

$$\mathbf{Y}_{TLBMP} = \mathbf{X_2} \cdot \mathbf{C} \qquad (6.33)$$

where

$$\mathbf{X}_{in} = \begin{bmatrix} x_{in}(n) & x_{in}(n)|x_{in}(n)| & \cdots & x_{in}(n)|x_{in}(n)|^{K_1-1} \\ x_{in}(n-1) & x_{in}(n-1)|x_{in}(n-1)| & \cdots & x_{in}(n-1)|x_{in}(n-1)|^{K_1-1} \\ \vdots & \vdots & \vdots & \vdots \\ x_{in}(n-N) & x_{in}(n-N)|x_{in}(n-N)| & \cdots & x_{in}(n-N)|x_{in}(n-N)|^{K_1-1} \end{bmatrix} \qquad (6.34)$$

$$\mathbf{X}_1 = \begin{bmatrix} 1 & x_1(n) & \cdots & x_1(n)|x_1(n)|^{K_2-1} \\ 1 & x_1(n-1) & \cdots & x_1(n-1)|x_1(n-1)|^{K_2-1} \\ \vdots & \vdots & \vdots & \vdots \\ 1 & x_1(n-N) & \cdots & x_1(n-N)|x_1(n-N)|^{K_2-1} \end{bmatrix} \qquad (6.35)$$

$$\mathbf{X}_2 = \begin{bmatrix} x_2(n) & x_2(n)|x_2(n)|^{K_3-1} & \cdots & x_2(n-M)|x_2(n-M)|^{K_3-1} \\ x_2(n-1) & x_2(n-1)|x_2(n-1)|^{K_3-1} & \cdots & x_2(n-1-M)|x_2(n-1-M)|^{K_3-1} \\ \vdots & \vdots & \vdots & \vdots \\ x_2(n-N) & x_2(n-N)|x_2(n-N)|^{K_3-1} & \cdots & x_2(n-N-M)|x_2(n-N-M)|^{K_3-1} \end{bmatrix}$$

(6.36)

and

$$\mathbf{Y}_{TLBMP} = \begin{bmatrix} y_{TLBMP}(n) \\ y_{TLBMP}(n-1) \\ \vdots \\ y_{TLBMP}(n-N) \end{bmatrix}$$

(6.37)

where **A**, **B**, and **C** are the vectors of the coefficients for each block. Further details on the definitions and formulations of the matrices and vectors in these equations are provided in [12]. These coefficients can be obtained using the linear least squares method, similar to the PLUME model.

Experimental validation has been conducted for a Doherty PA and a class AB PA; and, the TLBMP model exhibited a better performance than the conventional MP and orthogonal MP models (Section 5.4.1), while reducing the complexity significantly in terms of the number of coefficients and the number of operations to compute the models.

6.6.3 Rational Function Model for Amplifiers

Another important polynomial model that can be used for compensation of PA nonlinearity is the rational function model. Rational functions are universal approximators and can be used for the estimation and detection of signals, such as radar signals [13]. A rational polynomial is the ratio of two polynomials given by:

$$y_{RF}(n) = \frac{\sum\limits_{j=0}^{J} a_j \cdot x^j(n)}{\sum\limits_{k=0}^{K} b_k \cdot x^k(n)}.$$

(6.38)

Here, J and K are the nonlinearity orders of the numerator and denominator, respectively. An absolute-term denominator rational function (ADRF) is given by [14]:

$$y_{ADRF}(n) = \frac{\sum\limits_{m_1=0}^{M_1} \sum\limits_{j=0}^{J} a_{m_1 j} \cdot x(n-m_1) \cdot |x(n-m_1)|^{2j}}{1 + \sum\limits_{m_2=0}^{M_2} \sum\limits_{k=0}^{K} b_{m_2 k} \cdot x(n-m_2) \cdot |x(n-m_2)|^{2k+1}}$$

(6.39)

This model includes memory in the system represented by M_1, M_2, J, and K are the nonlinearity orders; and, $a_{m_1 j}$ and $b_{m_2 k}$ are the coefficients of the model.

The method proposed in [14] uses a dynamic rational function (DRF) with a memoryless flexible order denominator (MFOD) as expressed in Equation 6.40. The DRF-MFOD model was shown to have the best performance compared to the conventional MP based model and the ADRF model [15]. In addition, the DFR-MFOD model described by Equation 6.40 is less complex and has fewer number of parameters than the ADRF model.

$$y_{DRF\_MFOD}(n) = \frac{\sum\limits_{m=0}^{M} \sum\limits_{j=0}^{J} a_{mj} \cdot x(n-m) \cdot |x(n-m)|^j}{1 + \sum\limits_{k=0}^{K} b_k \cdot x(n) \cdot |x(n)|^k}. \tag{6.40}$$

Here, M represents the memory depth, a_{mj} and b_k are the coefficients of the model; and, J and K are the nonlinearity orders of the numerator and denominator, respectively.

6.7 Polynomial Based Model with I/Q and DC Impairments

The previous sections focus mainly on the issue of PA nonlinearity. However, there are other imperfections related to various components of a transmitter. As mentioned earlier, the output, $y_{PA}(n)$, of the PA for a given input, $x_{in}(n)$, can be given by:

$$y_{PA}(n) = \sum\limits_{m=0}^{M} \sum\limits_{k=1}^{K} a_{mk} \cdot x_{in}(n-m) \cdot |x_{in}(n-m)|^{k-1}. \tag{6.41}$$

Here, M represents the memory depth and k is the nonlinearity index.

In actual transmitters, in addition to the nonlinearity caused by the PA, there are other problems that affect its performance. One such issue is the in-phase (I) and quadrature phase (Q) imbalance caused during the up-conversion of the input signal, giving rise to mirror frequency imaging and DC offset mainly due to carrier leakage [15]. Mathematically, the output of an I/Q modulator is given by [15]:

$$y_{I/Q}(n) = \sum\limits_{m=0}^{M} a_m \cdot x_{in}(n-m) + \sum\limits_{m=0}^{M} a_m \cdot x_{in}^*(n-m) + dc \tag{6.42}$$

where M is the memory present in the system and dc represents the dc offset. The second term in Equation 6.42 represents the image caused by the imbalance.

There are various methods that deal with imperfections in transmitters; two of which – the parallel Hammerstein based model and a generalized two-box model – are discussed in the following subsections.

There are other methods that mitigate various imperfections of transmitters. The Volterra series has been used to consider the effect of I/Q imbalance [16]. However, due to the large number of coefficients, the model bears very high complexity. A rational function based model for the joint alleviation of PA nonlinearity effects and I/Q imbalance has been proposed in [17].

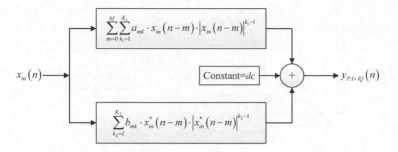

Figure 6.11   Parallel Hammerstein based model for the alleviation of PA nonlinearity and I/Q imbalance

6.7.1   Parallel Hammerstein (PH) Based Model for the Alleviation of Various Imperfections in Direct Conversion Transmitters

The parallel Hammerstein based model proposed in [18], shown in Figure 6.11, is a joint model that compensates for various imperfections in a transmitter. Prior to this model, a serial configuration based architecture that considered the effect of I/Q imbalance was presented. However, the drawback of the model was that the PA compensation and I/Q impairment compensation had to be processed separately. The joint model [18] converts this serial architecture into a parallel configuration using indirect learning architecture for parameter extraction, resulting in a single-step estimation. The serial-to-parallel conversion is detailed in [16, 18, 15]:

$$y_{PA+IQ}(n) = \sum_{m=0}^{M} \sum_{k_1=1}^{K_1} a_{mk} \cdot x_{in}(n-m) \cdot |x_{in}(n-m)|^{k_1-1}$$

$$+ \sum_{k_2=1}^{K_2} b_{mk} \cdot x_{in}^*(n-m) \cdot |x_{in}^*(n-m)|^{k_2-1} + dc. \qquad (6.43)$$

Here, M represents the memory index, k_1 and k_2 are the nonlinearity indices, respectively; and dc represents the local oscillator leakage.

6.7.2   Two-Box Model with I/Q and DC Impairments

A generalized two-box model, inspired by the FTNTB model, was proposed in [15] for the mitigation of PA distortions and I/Q modulator imperfections. The first block is composed of dual parallel branches of Volterra series, while the second block is a static nonlinear function. The block diagram of the two-box model is shown in Figure 6.12.

$x_{in}(n)$ is the input to the system, $x(n)$ is the output of the nonlinear FIR filters, while $y_{PA+IQ}(n)$ is the final output of the system. The dual parallel branch Volterra series consists of only the second order cross-terms and is mathematically represented as:

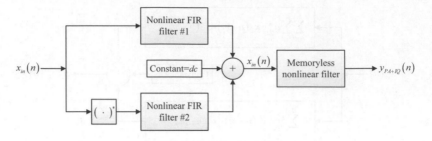

Figure 6.12   Two-box model with I/Q and DC impairments

$$x(n) = \left[\sum_{m=0}^{M} a_m \cdot x_{in}(n-m) + \sum_{m=0}^{M} \sum_{j=0}^{J} a_{mj} \cdot x_{in}(n-m) \cdot |x_{in}(n-m-j)| \right]$$

$$+ \left[\sum_{m=0}^{M} b_m \cdot x_{in}^*(n-m) + \sum_{m=0}^{M} \sum_{j=0}^{J} b_{mj} \cdot x_{in}^*(n-m) \cdot |x_{in}(n-m-j)| \right] + dc$$

$$\text{(6.44)}$$

where a_m, b_m, a_{mj}, and b_{mj} represent the model coefficients, M is the memory depth of the system, and J is the time delay of the envelope of the input signal $|x(\cdot)|$. The final output of the model is given by:

$$y_{PA+IQ}(n) = x(n)H(|x(n)|) \tag{6.45}$$

where $H(\cdot)$ is the complex static/memoryless gain of the LUT model. The coefficients of the model can be obtained using the least squares approach.

The two-box model reduces the complexity of the system more than the parallel Hammerstein model.

References

[1] Ghannouchi, F.M. and Hammi, O. (2009) Behavioural modeling and predistortion. *IEEE Microwave Magazine*, **10** (7), 52–64.
[2] Ghannouchi, F.M., Younes, M. and Rawat, M. (2013) Distortion and impairments mitigation and compensation of single and multi-band wireless transmitters. *IET Microwaves Antennas and Propagation*, **7** (7), 518–534.
[3] Liu, T., Boumaiza, S. and Ghannouchi, F.M. (2005) Deembedding static nonlinearities and accurately identifying and modeling memory effects in wide-band RF transmitters. *IEEE Transactions on Microwave Theory and Techniques*, **53** (11), 3578–3587.
[4] Haykin, S. (1996) *Adaptive Filter Theory*, 3rd edn, Prentice-Hall, Upper Saddle River, NJ.
[5] Liu, T., Boumaiza, S. and Ghannouchi, F.M. (2006) Augmented Hammerstein predistorter for linearization of broad-band wireless transmitters. *IEEE Transactions on Microwave Theory and Techniques*, **54** (4), 1340–1349.
[6] Taringou, F., Hammi, O., Srinivasan, B. *et al.* (2010) Behavior modeling of wideband RF transmitters using Hammerstein–Wiener models. *IET Circuits, Devices & Systems*, **4** (4), 282–290.

[7] Voros, J. (2007) An iterative method for Wiener–Hammerstein systems parameter identification. *Journal of Electrical Engineering*, **58** (2), 114–117.

[8] Olsson, D.M. and Nelson, L.S. (1975) The Nelder–Mead simplex procedure for function minimization. *Technometrics*, **17** (1), 45–51.

[9] Younes, M. and Ghannouchi, F.M. (2012) An accurate predistorter based on a feedforward Hammerstein structure. *IEEE Transactions on Broadcasting*, **58** (3), 454–461.

[10] O. Hammi and F. M. Ghannouchi, Twin nonlinear two-box models for power amplifiers and transmitters exhibiting memory effects with application to digital predistortion, *IEEE Microwave and Wireless Components Letters*, **19**, 8, 530–532, Aug. 2009.

[11] Younes, M., Hammi, O., Kwan, A. and Ghannouchi, F.M. (2010) An accurate complexity-reduced "PLUME" model for behavioral modeling and digital predistortion of RF power amplifiers. *IEEE Transactions on Industrial Electronics*, **58** (4), 1397–1405.

[12] Rawat, M., Ghannouchi, F.M. and Rawat, K. (2013) Three-layered biased memory polynomial for dynamic modeling and predistortion of transmitters with memory. *IEEE Transactions on Circuits and Systems I: Regular Papers*, **60** (3), 768–777.

[13] Leung, H. and Haykin, S. (1994) Detection and estimation using an adaptive rational function filter. *IEEE Transactions on Signal Processing*, **42** (12), 3366–3376.

[14] Rawat, M., Rawat, K., Ghannouchi, F.M. *et al.* (2014) Generalized rational functions for reduced-complexity behavioral modeling and digital predistortion of broadband wireless transmitters. *IEEE Transactions on Instrumentation and Measurement*, **63** (2), 485–498.

[15] Younes, M. and Ghannouchi, F.M. (2014) A generalised twin-box model for compensation of transmitters RF impairments. *IET Communications*, **8** (4), 413–418.

[16] Cao, H., Tahrani, A.S., Fager, C. *et al.* (2009) I/Q imbalance compensation using a nonlinear modeling approach. *IEEE Transactions on Microwave Theory and Techniques*, **57** (3), 513–518.

[17] Aziz, M., Rawat, M. and Ghannouchi, F.M. (2013) Rational function based model for the joint mitigation of I/Q imbalance and PA nonlinearity. *IEEE Microwave and Wireless Components Letters*, **23** (4), 196–198.

[18] Anttila, L., Handel, P. and Valkama, M. (2010) Joint mitigation of power amplifier and I/Q modulator impairments in broadband direct-conversion transmitters. *IEEE Transactions on Microwave Theory and Techniques*, **58** (4), 730–739.

7

Neural Network Based Models

7.1 Introduction

Recently, the technique of artificial neural networks (NNs) has drawn the attention of researchers in the field of power amplifier (PA) modeling due to its successful implementation and favorable results in pattern recognition, signal processing, system identification, and control [1–4]. As a result of its adaptive nature and its universal approximation capability [5–9], the NN approach has been investigated as one of the modeling and predistortion techniques for PAs and transmitters [5–14]. Different NN topologies and training algorithms that also take into account memory effects have been proposed. NN models can be either static (i.e., memoryless) or dynamic. A static NN model can be augmented to take into account memory effects and derive a dynamic model suitable for broadband nonlinear transmitters; such a model is often designated as a time-delay neural network (TDNN) [1, 3, 14].

This chapter presents major static and dynamic NN-based behavioral models and offers a comparison of different NN topologies and training algorithms that can be used for the identification of both forward and reverse models of nonlinear PAs and transmitters.

7.2 Basics of Neural Networks

NNs have been widely used as powerful tools for modeling nonlinear dynamic systems [15, 16]. The application of NNs to system modeling and identification is motivated by their universal approximation property, where a feedforward network with a finite number of neurons in a single hidden layer functions as universal approximator with a predetermined activation function. Like biological neural structures, the most basic component of a NN is the neuron. Each neuron consists of one output and one or multiple inputs. Each input is multiplied by a weight before entering the corresponding neuron. The weighted inputs are combined with reference to a bias/threshold

Behavioral Modeling and Predistortion of Wideband Wireless Transmitters, First Edition.
Fadhel M. Ghannouchi, Oualid Hammi and Mohamed Helaoui.

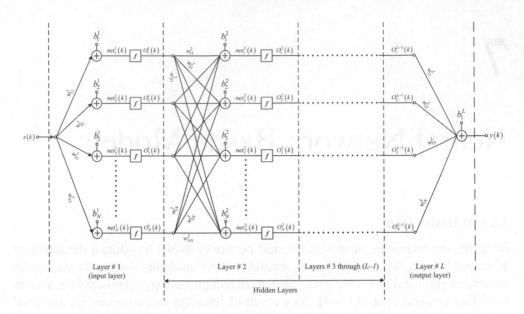

Figure 7.1 Block diagram of a typical L-layer feedforward neural network

value to calculate an intermediary output and the output of the neuron is calculated by applying an activation function to the intermediary output. Usually a sigmoid function is employed as an activation function in NNs.

Each set of neurons with the corresponding weights, bias, and activation function forms a layer. A layer receives all its inputs either from a preceding layer or directly from the external input in the case of the first layer. The output of a given neuron excites the succeeding layer's neurons as input (after modified by weights) or exits the NN as the final output to the external environment. If multiple layers are connected one after another, the intermediate layers that are between the input layer and the output layer are called the "hidden layers."

Recently, multilayer NNs, also called feedforward neural networks (FFNNs), have been used for modeling nonlinear memoryless transceivers and radio channels, such as traveling wave tube amplifiers (TWTAs) [6]. A typical L-layer and N-neuron FFNN is shown in Figure 7.1. In this figure, w_{ij}^l represents the synaptic weight between the output of neuron j at layer $(l-1)$ and the input of neuron i at layer l; net_i^l is the output value of neuron i at layer l after applying the bias/threshold b_i^l; and, O_i^l is the value of the output of neuron i at layer l after the application of activation function f.

In the FFNN model, the first layer has a single input that corresponds to the input signal $x(k)$ and N outputs. Conversely, the output layer has N inputs and a single output that represents the NN's output $y(k)$. Intermediate layers have each N inputs and N outputs. The jth output of the lth layer (O_j^l) is modified by a set of weights w_{ij}^{l+1} to generate the jth contribution at the input of the ith neuron of the $(l+1)th$ layer as depicted in Figure 7.1. All N inputs of the ith neuron of the $(l+1)th$ layer are combined along with a bias b_i^{l+1} to generate the *kth* sample of the output of the *ith* neuron of the $(l+1)th$ layer $[net_i^{l+1}(k)]$. This neuron's output is applied at the input of the activation function (f) to generate the *kth* sample of the output of the *ith* neuron of the $(l+1)th$ layer $[O_i^{l+1}(k)]$. The output of each layer is again modified by another set of weights before propagating to the next layer or feeding the final output.

For the FFNN in Figure 7.1, the operation of the NN can be summarized for the *ith* neuron $(i = 1, \cdots, N)$ of the *lth* layer as:

$$O_i^l(k) = f[net_i^l(k)] \quad \text{for } l = 1, \cdots, L-1 \tag{7.1}$$

where

$$\begin{cases} net_i^1(k) = w_{i1}^1 \cdot x(k) + b_i^1 \\[2ex] net_i^l(k) = \sum_{j=1}^{N} w_{ij}^l \cdot O_j^{l-1}(k) + b_i^l \quad \text{for } l = 2, \cdots, L-1 \end{cases} \tag{7.2}$$

The output of the FFNN becomes:

$$y(k) = \sum_{j=1}^{N} w_{1j}^L \cdot O_j^{L-1}(k) + b_1^L. \tag{7.3}$$

In Equation 7.1, the function f represents the activation function of the NN model. Common activation functions include the tan-sigmoid hyperbolic tangent, the log-sigmoid, Gaussian, and linear. These functions are illustrated in Figure 7.2.

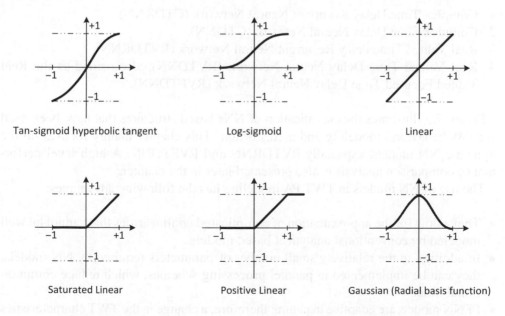

Figure 7.2   Most common activation functions used in artificial neural network

The standard back-propagation training algorithm [1] can be employed to train the FFNN. Once an FFNN is trained, it becomes the model of the system under consideration. FFNN models have been shown to outperform the classic analytical and behavioral models discussed in Chapter 4 for TWT PAs [17, 18]. The NN architectures for modeling nonlinear behavior of PAs can be categorized into two major groups depending on the existence of memory in the network:

1. Static NN models.
2. Dynamic NN models.

As the names suggest, static models characterize the static or memoryless nonlinearity, whereas the dynamic nonlinearity or the nonlinearity with memory is characterized by dynamic NN models. The complex static NN models can be categorized in three main architecture classes:

1. Single-Input Single-Output Feedforward Neural Network (SISO-FFNN)
2. Dual-Input Dual-Output Feedforward Neural Network (DIDO-FFNN) or Real valued neural network (RVNN).
3. Dual-Input Dual-Output Coupled Cartesian based Neural Network (DIDO-CC-NN)

Similarly, complex dynamic NN models can be classified in three main architecture classes as follows:

1. Complex Time Delay Recurrent Neural Network (CTDRNN).
2. Complex Time Delay Neural Network (CTDNN).
3. Real Valued Time delay Recurrent Neural Network (RVTDRNN).
4. Real Valued Time Delay Neural Network (RVTDNN), also named as the Real Valued Focused Time Delay Neural Network (RVFTDNN).

Figure 7.3 illustrates the classification of NNs based structures that have been used for PAs behavioral modeling and predistortion. This chapter mainly focuses on the dynamic NN models, especially RVTDRNN and RVFTDNN. A high level performance comparison analysis is also presented later in the chapter.

The use of NN models in TWT PA modeling has the following advantages:

- They allow for the approximation of complicated nonlinearities that cannot be well modeled by conventional analytical based models.
- In addition to the relatively small number of parameters required by NN models, they can be implemented in parallel processing schemes, which reduce computational time.
- FFNN models are adaptive in nature; therefore, a change in the TWT characteristics can be tracked relatively with ease.

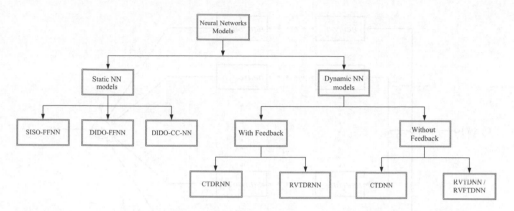

Figure 7.3   Classification of neural networks structures for power amplifiers modeling and predistortion applications

- NN models ensure good mathematical properties such as accurate asymptotic behavior approximation and ensure continuity of response.

7.3   Neural Networks Architecture for Modeling of Complex Static Systems

7.3.1   Single-Input Single-Output Feedforward Neural Network (SISO-FFNN)

To model complex static systems, three NN-based architectures can be adopted. The most basic structure proposed is a SISO-FFNN utilizing complex input/output signals [6], as illustrated in Figure 7.4. In this architecture, a complex input is modified with a set of complex valued weights before entering the input layer. The output of each neuron goes to the next layer of neurons and modified again with another set of complex weights and biases before it gets added in the final layer to obtain final output. Each neuron $neuron_i^l$ consists of complex input, complex weights, bias, and activation function. This architecture introduces complex valued weights and activation functions, which usually result in cumbersome calculations and divergence when training the network. The input-output relationship of this model can be expressed as:

$$C_{out}(k) = \sum_{j=1}^{N} z_{1j}^{L} \cdot O_j^{L-1}(k) + b_1^{L} \qquad (7.4)$$

where the output of the of any neuron at an intermediate layer can be calculated using the same scheme described by Equations 7.1 and 7.2. The main difference here is that the synaptic weights (z_{ij}^l), the biases (b_i^l), as well as the activation functions (f), are complex valued.

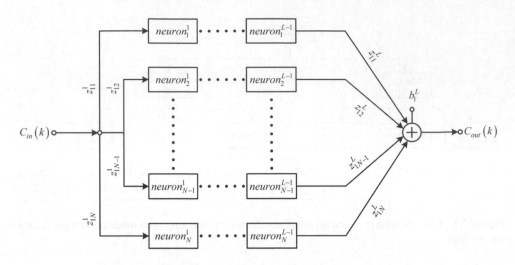

Figure 7.4   Single-input single-output feedforward neural network

7.3.2   Dual-Input Dual-Output Feedforward Neural Network (DIDO-FFNN)

The second architecture proposed to model complex static systems is based on splitting the complex data stream into two components, which are then processed separately using two real-valued feedforward neural networks (RVFFNNs) as shown in Figure 7.5 and proposed in [7] and [8]. The RVFFNN is similar to a typical complex FFNN but only takes real values for inputs. In the polar based architecture, the complex input signal C_{in} is first decomposed into its polar (A_{in}, Φ_{in}) components. In architecture, the magnitude A_{in} and phase Φ_{in} of the input signal feed the input of the first and second real valued NNs, respectively. Similarly, in the Cartesian based architecture of this model, the signal is first decomposed into its Cartesian components (I_{in}, Q_{in}). Then, the real or in-phase I_{in} and imaginary or quadrature-phase Q_{in} parts of the input signal feed the input of the first and second real-valued NNs, respectively. Finally, the outputs of both real-valued FFNNs are recombined to construct the complex output signal. Thus, this can be seen as a dual-input dual-output (DIDO) polar or Cartesian based architecture.

The DIDO polar based FFNN architecture utilizes two uncoupled NNs that attempt to capture the AM/AM (amplitude modulation to amplitude modulation) and AM/PM (amplitude modulation to phase modulation) responses separately. The main drawback of this topology is the asynchronous convergence of the separate phase and amplitude FFNNs, where both NNs do not converge to an optimal model at same time, leading to over- or under-training of one NN. In the Cartesian based DIDO-FFNN architecture, separate real valued FFNNs are used to model the in-phase (I) and quadrature-phase (Q) components of the system's output signal. This approach takes advantage of the availability of the I and Q components, but it is also prone to asynchronous convergence between the I and Q sub-models of the NN.

First real-valued neural network for magnitude modeling

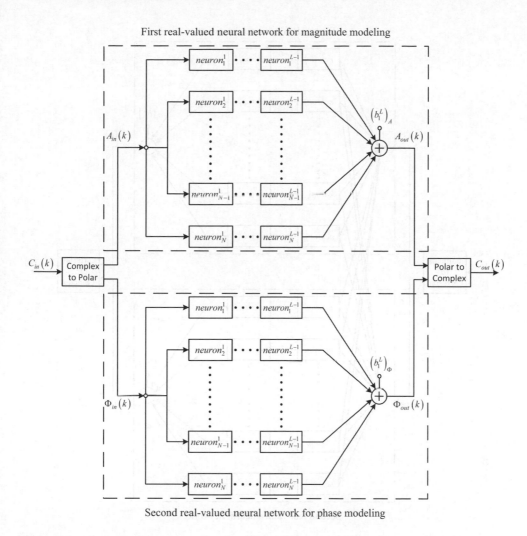

Second real-valued neural network for phase modeling

Figure 7.5   Block diagram of dual-input dual-output feedforward neural network applied on the polar components

7.3.3   Dual-Input Dual-Output Coupled Cartesian Based Neural Network (DIDO-CC-NN)

To avoid asynchronous convergence of the DIDO polar or Cartesian based FFNN, the DIDO-CC-NN was proposed in [14]. As depicted in Figure 7.6, the structure of this model decomposes the complex input signal C_{in} into its Cartesian components (I_{in}, Q_{in}), which are then simultaneously fed to two separate RVFFNNs. The major difference with the previous models is that in this case, both Cartesian components in the input layer are coupled with both NNs. Finally, the outputs of two RVFFNNs are recombined to obtain the complex output signal.

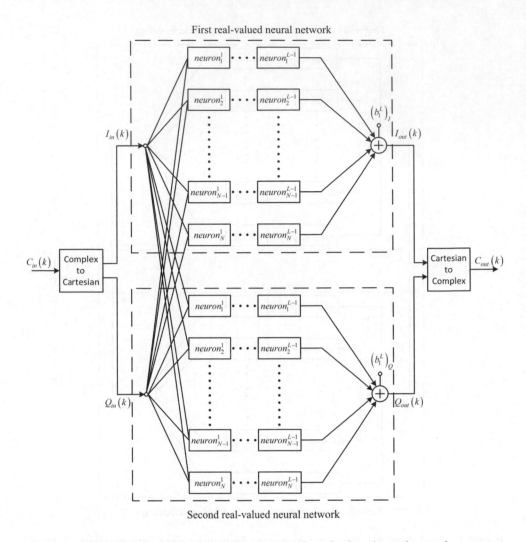

Figure 7.6   Dual-input dual-output coupled Cartesian based neural network

The three architectures described in this section have been found to be effective, to various extents, for the forward modeling of systems with strong static nonlinearity; they all fall short of expectations when the system exhibits strong dynamics and memory effects.

7.4   Neural Networks Architecture for Modeling of Complex Dynamic Systems

To account for the presence of memory effects, dynamic neural structures have been proposed in the literature. These dynamic NNs structures can be sub-divided into two categories according to the nature of their architecture:

1. NNs architectures without feedback such as the CTDNNs, and RVFTDNNs.
2. NNs architectures with feedback such as CTDRNNs and RVTDRNNs.

These major dynamic models are discussed in the following subsections.

7.4.1   Complex Time-Delay Recurrent Neural Network (CTDRNN)

One of the most popular NN models is the CTDRNN model, which utilizes feedforward and feedback signal propagating schemes [5, 14]. An illustration of a typical
feedback CTDRNN is shown in Figure 7.7.

  In this architecture, the input signal is fed to the input layer through a set tapped
delay lines (TDLs) containing p branches. A tapped delayed feedback of the output is
also fed to the input layer. The feedback path TDLs are made of q branches. Both input
sets, including the $(p + 1)$ delayed samples associated with the input signal $x(k)$ and
the q delayed samples of the output signal $y(k)$, are fed to the first layer's N neurons
using synaptic weights and biases. An activation function is applied and the outputs
of each neuron are again scaled with complex weights. Finally the weighted outputs
are added together to obtain the final output.

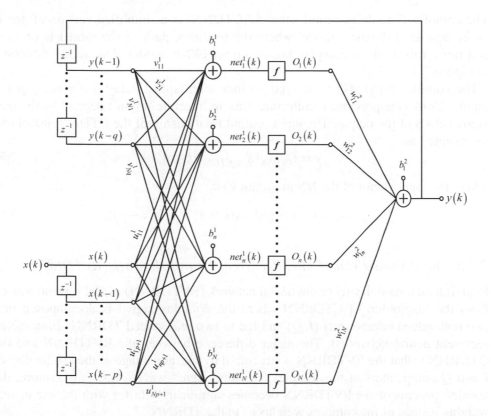

Figure 7.7   Block diagram of a two-layer complex time-delay recurrent neural network (CTDRNN)

CTDRNNs use single-input single-output (SISO) complex architecture and hence suffer from cumbersome calculations and divergence when training the network. The dynamics of the system and memory effects are considered by taking into account the previous input and output samples, $x(k-1)$ through $x(k-p)$ and $y(k-1)$ to $y(k-q)$, respectively; where p and q represent the memory depth of the system. The input–output relationship of the CTDRNN model is given as:

$$y = f_{CTDRNN}[X_{in\_CTDRNN}(k, p, q)] \tag{7.5}$$

where the input vector of the NN at instant k is:

$$X_{in\_CTDRNN}(k, p, q) = [x(k), x(k-1), \cdots, x(k-p), y(k-1), y(k-2), \cdots, y(k-q)] \tag{7.6}$$

Thus the network can be seen as having $(p + 1 + q)$ complex inputs. Furthermore, the addition of the feedback delay between the inputs and outputs of the network increases the computational complexity and, often, negatively impacts the training and convergence of the network.

7.4.2 Complex Time-Delay Neural Network (CTDNN)

The complex time-delay neural network (CTDNN) is a simplified version of previously described dynamic model where the feedback path of the model is omitted and make this model a complex non-recurrent FFNN model. This model is shown in Figure 7.8.

The complex input $x(k)$ is delayed p times with tapped delay line making $p + 1$ inputs. Unlike the previous architecture, this architecture doesn't depend on the previous values of the output. The input–output relationship of the CTDNN model can be obtained as:

$$y = f_{CTDNN}[X_{in\_CTDNN}(k, p)] \tag{7.7}$$

where the input vector of the NN at instant k is:

$$X_{in\_CTDNN}(k, p) = [x(k), x(k-1), \cdots, x(k-p)]. \tag{7.8}$$

7.4.3 Real Valued Time-Delay Recurrent Neural Network (RVTDRNN)

Real-valued time-delay recurrent neural network (RVTDRNN) model was introduced from the inspiration of CTDRNN where the complex signal is decomposed into two real valued components (I, Q) and fed to two real valued TDRNNs (time-delay recurrent neural networks). The major difference between the RVTDRNN and the CTDRNN is that the RVTDRNN is structured to take advantage of the availability of I and Q components of the complex baseband signal waveforms. Furthermore, the training process of the RVTDRNN becomes significantly faster with the use of real weights instead of the complex weights as in the TDRNN.

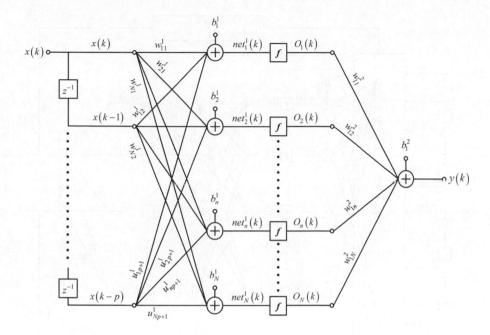

Figure 7.8 Block diagram of complex time delay neural network (CTDNN)

In this architecture, shown in Figure 7.9, the input and output signals' in-phase and quadrature components, (I_{in}, Q_{in}) and (I_{out}, Q_{out}), are delayed with TDLs for p and q times, respectively. Therefore, the order of the input vector for RVTDRNN at any moment of the training sequence is $2(p+1+q)$-by-1 including past samples of the input and output signals [12]. Here, p and q are the memory orders of the input and feedback signals, respectively. The input expression is given as,

$$
\begin{aligned}
&X_{in\_RVTDRNN}(k,p,q) \\
&= \left[\begin{array}{l} I_{in}(k), I_{in}(k-1), \cdots, I_{in}(k-p), Q_{in}(k), Q_{in}(k-1), \cdots, Q_{in}(k-p) \\ I_{out}(k-1), I_{out}(k-2), \cdots, I_{out}(k-q), Q_{out}(k-1), Q_{out}(k-2), \cdots, Q_{out}(k-q) \end{array} \right]
\end{aligned}
$$
$$(7.9)$$

The in-phase and quadrature-phase components of the output signal, I_{out} and Q_{out}, respectively; are given by:

$$I_{out}(k) = f_I[X_{in\_RVTDRNN}(k,p,q)] \qquad (7.10)$$

$$Q_{out}(k) = f_Q[X_{in\_RVTDRNN}(k,p,q)] \qquad (7.11)$$

where f_1 and f_2 are activation functions modeled by RVRNN (real valued recurrent neural network).

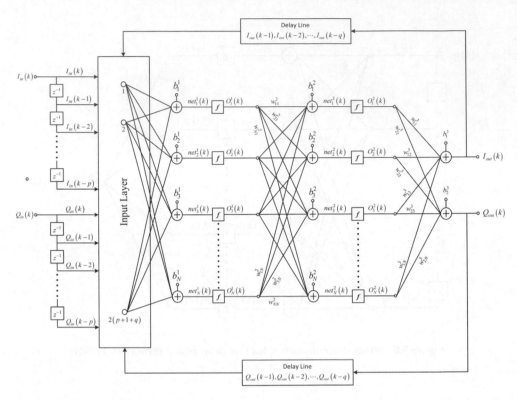

Figure 7.9 Block diagram of a three-layer real valued time delay recurrent neural network (RVTDRNN)

7.4.4   Real Valued Time-Delay Neural Network (RVTDNN)

This model is obtained through adding TDLs to the FFNN structure. The RVTDNN was proposed in [14] and has been found to be effective in modeling strongly dynamic nonlinear systems, such as wideband PAs and wireless transmitters. This model has also been designated as a real value focused time-delay neural network (RVFTDNN) [12]. As illustrated in Figure 7.10, RVFTDNN architecture is based on the previously discussed static DIDO-CC-NN architecture and taking further into account the memory effects and assuming that the output of the amplifier depends only on the present and $2p$ past input values, but not on the network's output values.

Figure 7.10 shows a three-layer RVFTDNN with two real inputs (I_{in} and Q_{in}), and two real outputs (I_{out} and Q_{out}). The inputs I_{in} and Q_{in} are both delayed by p samples; using two sets of TDLs. The first set of TDLs is made of p branches and is applied on the in-phase component of the input signal (I_{in}), while the second set of TDLs is also made of p branches and is applied to the quadrature-phase component of the input signal (Q_{in}). Here, p represents the memory depths of the system and the length of the input vectors are $2(p+1)$-by-1. The delayed response is achieved by using a plurality

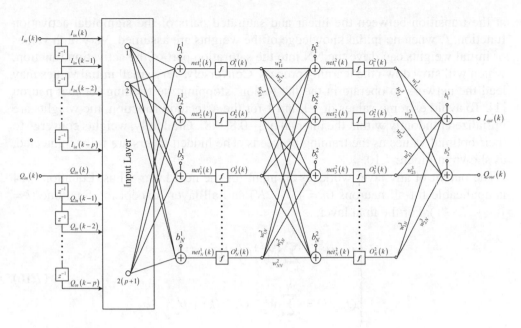

Figure 7.10   Block diagram of a three-layer real valued focused time-delay neural network

of unit delay operators(z^{-1}), where the unit delay operator yields to the delayed sample $x(k-1)$ when operating on sample $x(k)$. The input-output relationship for this model with n neurons and two layers can be written as,

$$I_{out}(k) = f_I[I_{in}(k), I_{in}(k-1), \cdots, I_{in}(k-p), Q_{in}(k), Q_{in}(k-1), \cdots, Q_{in}(k-p)] \tag{7.12}$$

$$Q_{out}(k) = f_Q[I_{in}(k), I_{in}(k-1), \cdots, I_{in}(k-p), Q_{in}(k), Q_{in}(k-1), \cdots, Q_{in}(k-p)] \tag{7.13}$$

At any time instant k, the output value of neuron n of layer l is given by:

$$net_n^l(k) = \sum_{j=1}^{N} w_{nj}^l O_j^{l-1}(k) + b_n^l \tag{7.14}$$

where w_{nj}^l is the synaptic weight between the output of neuron j at layer $(l-1)$ and the input of neuron n at layer l, b_n^l refers to the bias applied to the neuron n at layer l, and $O_j^{l-1}(k)$ is the output, at time instant k, of the neuron and j at layer $(l-1)$. $O_j^{l-1}(k)$ is given by:

$$O_j^{l-1}(k) = f[net_j^{l-1}(k)]. \tag{7.15}$$

The synaptic weights w_{ij}^l, between the output of neuron j at layer $(l-1)$ and the input of neuron i at layer l, are chosen such that the output values of all neurons lie at the transition between the linear and saturated parts of the sigmoidal activation function, f, when no initial knowledge of the weights are assumed. Very high values of initial weights can drive the NN into the saturation part of the activation function, which will slow down the learning process. Conversely, very small initial values may lead the network to operate in the flat region, stopping the training for that neuron [1]. To avoid extreme values of -1 and 1 for the activation function, the weights are initialized randomly within the interval of $[-0.8, 0.8]$. Gradually, weights converge to their optimal values as the training proceeds. The hidden layers are fully connected, as shown in Figure 7.10.

The output of any layer works as an input for the next layer. Thus, Equation 7.14 is applicable for all neurons $(n = 1, \cdots, N)$ and all layers except the final one $(l = 1, \cdots, L - 1)$. For the final layer,

$$
\begin{cases}
I_{out}(k) = \displaystyle\sum_{j=1}^{N} w_{1j}^{L-1} O_j^{L-1}(k) + b_1^l \\[2mm]
Q_{out}(k) = \displaystyle\sum_{j=1}^{N} w_{2j}^{L-1} O_j^{L-1}(k) + b_2^l.
\end{cases}
\tag{7.16}
$$

The output layer has a purely linear activation function (sometimes referred to as a *purelin* function), which sums up the outputs of hidden neurons and linearly maps them to the output. The activation function, f, for the two hidden layers can be chosen as one of the functions depicted in Figure 7.2. It is worth noting that when the memory depth is zero (i.e., $p = 0$), the two TDLs are eliminated and in this case the RVFTDNN architecture is reduced to a RVFFNN architecture.

Training is carried out in batch modes, supervised with a back-propagation algorithm. Detailed descriptions of the back-propagation algorithm are given in [1, 16, 19, 20]. To summarize, two passes are made during one ensemble of iterations (commonly referred to as an epoch which can include tens to hundreds of actual iterations): a forward propagation and a backward propagation. During the forward propagation, the cost function is calculated by:

$$
E = \frac{1}{2K} \sum_{k=1}^{K} \{ [I_{out}(k) - \hat{I}_{out}(k)]^2 + [Q_{out}(k) - \hat{Q}_{out}(k)]^2 \}
\tag{7.17}
$$

where $I_{out}(k)$ and $Q_{out}(k)$ are the desired model outputs representing the Cartesian components of the system's complex output; and $\hat{I}_{out}(k)$ and $\hat{Q}_{out}(k)$ are their predicted values by the actual NN model.

Based on the error signal given by Equation 7.17, a backward computation is performed to adjust the synaptic weights of the network in layer l according to:

$$
w_{nj}^l(k + 1) = w_{nj}^l(k) + \Delta w_{nj}^l(k)
\tag{7.18}
$$

In Equation 7.18, $w_{nj}^l(k)$ and $w_{nj}^l(k + 1)$ denote the values of the synaptic weight w_{nj}^l during the training step at instants k and $(k + 1)$, respectively; and $\Delta w_{nj}^l(k)$ is the adjustment applied to modify the value of the synaptic weight $w_{nj}^l(k)$ to obtain its new value at instant $(k + 1)$. $\Delta w_{nj}^l(k)$ is calculated, at instant k, using the one-dimensional Levenberg–Marquardt (LM) algorithm [20]. The LM algorithm was found to be the most appropriate among various algorithms for its fast convergence properties, as shown in the next section. The whole procedure is carried out until the desired per-formance is met, or the NN fails the validation procedure by drifting away from the generalization criterion [1, 16, 21, 22].

Due to their dynamic modeling capability, the RVRNN and RVFTDNN are consid-ered as good candidates among all the other NN topologies for the dynamic modeling and linearization based on digital predistortion techniques for PAs and transmitters.

In [12], the performance of the RVFTDNN was benchmarked against the RVRNN for a Doherty PA prototype driven by multi-carrier WCDMA signal. Each model had two hidden layers and the number of neurons in each of these layers was decided through the same optimization process. The output layer of each model contained two linear neurons. The ability of the models to predict the magnitude and phase of the output signal are reported in Figures 7.11 and 7.12, respectively. Both figures illustrate the superior performance of the RVFTDNN as it can accurately predict the output signal's magnitude and phase.

7.5   Training Algorithms

Another key factor for an NN is the learning algorithm used, which influences the speed, accuracy, and generalization during the learning process. Among the most pop-ular training algorithms that have been widely used for NNs, one can find:

- The Broyden–Fletcher–Goldfarb–Shanno quasi-Newton algorithm (BFGS).
- The Powell–Beale conjugate gradient algorithm (CGB).
- The Fletcher–Powell conjugate gradient algorithm (CGF).
- The Polak–Ribiere conjugate gradient algorithm (CGP).
- The gradient descent with adaptive learning rate algorithm (GDA).
- The gradient descent with momentum algorithm (GDM).
- The momentum and adaptive learning rule algorithm (GDX).
- The Levenberg–Marquardt algorithm (LM).
- The one-step secant algorithm (OSS).
- The resilient back-propagation algorithm (RP).
- The scaled conjugate gradient algorithm (SCG).

The GDA, GDX, GDM, and RP algorithms are first order optimization techniques that ignore second order and higher terms to provide a solution with less memory consumption for adaptive applications. Second order optimization techniques, such as BFGS, LM, and conjugate gradient techniques, take Hessian matrix information into account to achieve faster convergence, but have much higher memory requirements [21–25].

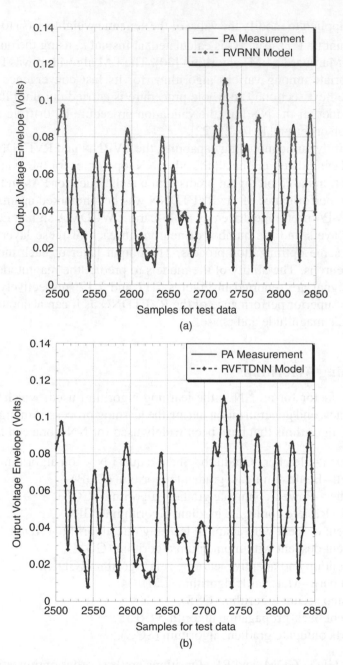

Figure 7.11 Predicted output voltage magnitude of a Doherty power amplifier prototype. (a) RVRNN model and (b) RVFTDNN model [12]. ©2009 IEEE. Reprinted, with permission, from Rawat et al., "Adaptive digital predistortion of wireless power amplifiers/transmitters using dynamic real-valued focused time delay line neural networks," *IEEE Transactions on Microwave Theory and Techniques*, Jan. 2010

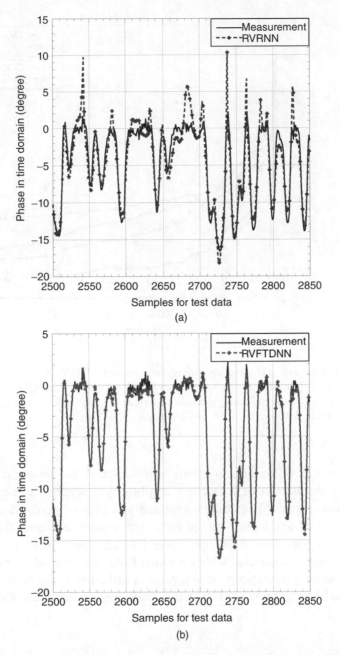

Figure 7.12   Predicted output voltage phase of a Doherty power amplifier prototype. (a) RVRNN model
and (b) RVFTDNN model [12]. ©2009 IEEE. Reprinted, with permission, from Rawat et al., "Adaptive
digital predistortion of wireless power amplifiers/transmitters using dynamic real-valued focused time
delay line neural networks," *IEEE Transactions on Microwave Theory and Techniques*, Jan. 2010

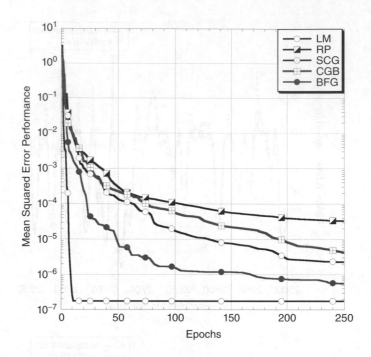

Figure 7.13 Comparison of training algorithms performance for RVFTDNN based digital predistorter [12]. ©2009 IEEE. Reprinted, with permission, from Rawat et al., "Adaptive digital predistortion of wireless power amplifiers/transmitters using dynamic real-valued focused time delay line neural networks," *IEEE Transactions on Microwave Theory and Techniques*, Jan. 2010

In [12], several of these algorithms were applied for to build a RVFTDNN based digital predistorter of a Doherty amplifier prototype driven by multi-carrier communication signals. The comparative analysis of their performance in terms of convergence speed and mean squared error performance is reported in Figure 7.13. The results illustrate the advantage of the LM algorithm in terms of speed as it converges in only few epochs, which is much faster than the other tested algorithms that require at least 100 epochs to converge. Furthermore, the plots in Figure 7.13 show that the LM algorithm leads to better modeling accuracy than the other learning algorithms.

7.6 Conclusion

In this chapter, a review of the NN-based models for behavioral modeling and digital predistortion of wireless transmitters was presented. Discussion of the various structures for NN models for addressing static and dynamic systems, their pros and cons and the appropriate implementation was presented. The performance of RVRNN and RVFTDNN architectures was compared using experimental results. It was shown that the RVFTDNN architecture is an appropriate architecture for addressing memory effects and nonlinear distortion of wireless transmitters. The RVRNN architecture was not able to outperform the RVFTDNN in spite of its complexity.

References

[1] Haykin, S. (1999) *Neural Networks: A Comprehensive Foundation*, Prentice-Hall, Upper Saddle River, NJ.

[2] Zhang, Q.J. and Gupta, K.C. (2000) *Neural Networks for RF and Microwave Design*, Artech House, Norwood, MA.

[3] Bengio, Y. (1995) *Neural Networks for Speech and Sequence Recognition*, ITC Press, New York, NY.

[4] Xu, J., Yagoub, M.C.E., Ding, R. and Zhang, Q.J. (2002) Neural-based dynamic modeling of nonlinear microwave circuits. *IEEE Transactions on Microwave Theory and Techniques*, **50** (12), 2769–2780.

[5] D. Luongvinh and Y. Kwon, Behavioral modeling of power amplifiers using fully recurrent neural networks. Digest 2005 IEEE MTT-S International Microwave Symposium (IMS), Long Beach, CA, June 2005, pp. 1979–1982, 2005.

[6] Ibnkahla, M., Sombrin, J., Castanie, F. and Bershad, N.J. (1997) Neural networks for modeling nonlinear memoryless communication channels. *IEEE Transactions on Communications*, **45** (7), 768–771.

[7] N. Benvenuto, F. Piazza, and A. Uncini, A neural network approach to data predistortion with memory in digital radio systems. Proceedings IEEE International Conference on Communications, Geneva, Switzerland, May 1993, pp. 232–236, 1993.

[8] N. Naskas and Y. Papananos, Adaptive baseband predistorter for radio frequency power amplifiers based on a multilayer perceptron. Proceedings 9th International Conference on Electronics, Circuits and Systems, Dubrovnik, Croatia, September 2002, pp. 1107–1110, 2002.

[9] Y. Quian and F. Liu, Neural network predistortion technique for nonlinear power amplifiers with memory. Proceedings First International Conference on Communications and Networking in China (ChinaCom'06), Beijing, China, October 2006, pp. 1–5, 2006.

[10] Rawat, M., Rawat, K., Younes, M. and Ghannouchi, F.M. (2013) Joint mitigation of non-linearity and modulator imperfections in a dual-band concurrent transmitter using neural networks. *Electronics Letters*, **49** (4), 253–255.

[11] Rawat, M. and Ghannouchi, F.M. (2012) A mutual distortion and impairment compensator for wideband direct conversion transmitters using neural networks. *IEEE Transactions on Broadcasting*, **58** (2), 168–177.

[12] Rawat, M., Rawat, K. and Ghannouchi, F.M. (2010) Adaptive digital predistortion of wireless power amplifiers/transmitters using dynamic real-valued focused time delay line neural networks. *IEEE Transactions on Microwave Theory and Techniques*, **58** (1), 95–104.

[13] Rawat, M. and Ghannouchi, F.M. (2012) Distributive spatiotemporal neural network for nonlinear dynamic transmitter modeling and adaptive digital predistortion. *IEEE Transactions on Instrumentation and Measurement*, **61** (3), 595–608.

[14] Liu, T., Boumaiza, S. and Ghannouchi, F.M. (2004) Dynamic behavioral modeling of 3G power amplifiers using real-valued time-delay neural networks. *IEEE Transactions on Microwave Theory and Techniques*, **52** (3), 1025–1033.

[15] Narendra, K.S. (1996) Neural networks for control theory and practice. *Proceedings of the IEEE*, **84** (10), 1385–1406.

[16] Narendra, K.S. and Parthasarathy, K. (1990) Identification and control of dynamical systems using neural networks. *IEEE Transactions on Neural Networks*, **1** (1), 4–27.

[17] A. Leke and J. S. Kenney, Behavioral modeling of narrowband microwave power amplifiers with applications in simulating spectral regrowth. Digest 1996 IEEE MTT-S International Microwave Symposium (IMS), San Francisco, CA, June 1996, pp. 1385–1388, 1996.

[18] A. Ghorbani and M. Sheikhan, The effect of solid state power amplifiers (SSPAs) nonlinearities on MPSK and M-QAM signal transmission. Proceedings Sixth International Conference on Digital Processing of Signals in Communications, Loughborough, UK, September 1991, pp. 193–197, 1991.

[19] R. Hecht-Nielsen, Theory of the back propagation neural network. Proceedings International Joint Conference on Neural Networks, San Diego, CA, June 1989, pp. 593–608, 1989.

[20] Hagan, M.T. and Menhai, M.B. (1994) Training feedforward network with the Marquardt algorithm. *IEEE Transactions on Neural Networks*, **5** (6), 989–993.

[21] E. M. L. Beale, A derivation of conjugate gradients, in *Numerical Methods for Nonlinear Optimization*, F. A. Lootsma (ed.). Academic Press, London, pp. 39–43, 1972.

[22] Battiti, R. (1992) First and second order methods for learning: Between steepest descent and Newton's method. *Neural Computation*, **4** (2), 141–166.

[23] Fletcher, R. and Reeves, C.M. (1964) Function minimization by conjugate gradients. *Computer Journal*, **7** (2), 149–154.

[24] Moller, M.F. (1993) A scaled conjugate gradient algorithm for fast supervised learning. *Neural Networks*, **6** (4), 525–533.

[25] Hagan, M.T., Demuth, H.B. and Beale, M.H. (1996) *Neural Network Design*, PWS Publishing, Boston, MA.

[26] Saleh, A.A.M. (1981) Frequency-independent and frequency-dependent nonlinear models of TWT amplifiers. *IEEE Transactions on Communications*, **29** (11), 1715–1720.

8

Characterization and Identification Techniques

8.1 Introduction

In the previous chapters, a thorough review of behavioral models proposed for the modeling and predistortion of wideband power amplifiers (PAs) and transmitters was presented. All these models can be seen as mathematical functions for which a set of coefficients needs to be identified. These coefficients are derived, using identification techniques, from measurements data acquired through the characterization of the device under test (DUT). Thus, the validity of a behavioral model and its accuracy will greatly depend, among others, on the characterization step. In fact, since the behavioral model coefficients are calculated solely from input and output measured data, the obtained model is able to take into consideration only the effects that are observed during the characterization step. For example, if measurements are performed with a test signal for which a DUT has a memoryless behavior, then a model derived from these measurements will be unable to predict the memory effects that will be present in the DUT if a wider bandwidth signal is used. This is true even if the model structure incorporates memory effects (such as the memory polynomial model). The accuracy of the model also depends on the model structure that is adopted and its ability to mimic all aspects of the observed behavior. As matter of fact, if the DUT exhibits memory effects during the measurements, the appropriate model structure should be used to ensure that the model reproduces these memory effects. For instance, a memoryless model cannot predict the memory effects of the DUT even if it is identified using measurements that include memory effects.

Accordingly, choosing the adequate model structure is certainly required but definitely not enough to guarantee accurate behavioral modeling and high performance digital predistortion (DPD). To better recognize the factors affecting the performance of a behavior model, the flow chart of behavioral modeling and DPD processes

Behavioral Modeling and Predistortion of Wideband Wireless Transmitters, First Edition.
Fadhel M. Ghannouchi, Oualid Hammi and Mohamed Helaoui.
© 2015 John Wiley & Sons, Ltd. Published 2015 by John Wiley & Sons, Ltd.

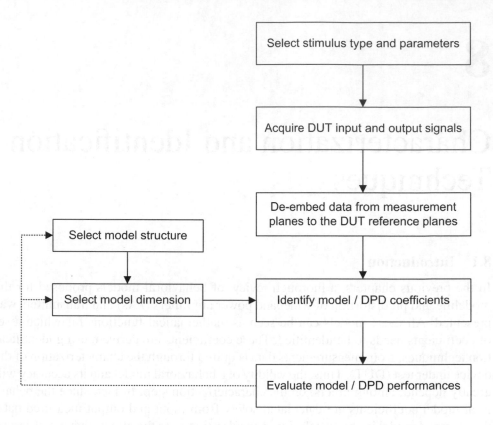

Figure 8.1   Flow chart of behavioral modeling and digital predistortion processes

is illustrated in Figure 8.1. First, the type of the drive signal to be used in the characterization of the DUT is selected. This encompasses continuous wave (CW), two-tones, multi-tones, as well as standard compliant and synthetic test signals. Then, the measurement data are acquired at the input and the output of the DUT. The raw measurements need to be processed to de-embed the signals from the measurement reference planes to the DUT reference planes. Next, the behavioral model and/or digital predistorter structure is selected, its parameters (i.e., nonlinearity order, memory depth, ...) defined, and its coefficients identified. Finally, the performance of the behavioral model and/or digital predistorter is assessed. The performance of a behavioral model is evaluated by comparing the predicted and the measured output signals of the DUT and quantifying the similarity between these two signals using the metrics defined in Chapter 3. The performance of a digital predistorter is quantified in frequency domain by the adjacent channel leakage ratio measured at the output of the linearized DUT, and in modulation domain by evaluating the error vector magnitude at the output of the linearized DUT. Based on the performance of the model/DPD, the model parameters can be adjusted, or if needed, a different model structure can be selected.

This chapter focuses on the types of test signals that can be used to characterize the behavior of power amplifiers and wireless transmitters, and the impact of these test signals on the observed behavior of the DUT. Emphasis is then given to the use of standard compliant test signals and the steps required for processing the measured data under such conditions, as well as the identification techniques that can be applied to calculate the model coefficients.

8.2   Test Signals for Power Amplifier and Transmitter Characterization

A variety of test signals have been proposed for the characterization of power amplifiers and wireless transmitters. Historically, continuous wave signals were initially used to derive the power dependent gain of the DUT. Later, two-tone and then multi-tone signals were applied. These test signals are more advanced than the continuous wave signal and present fertile ground for analytical derivations of power amplifiers' and transmitters' nonlinear behavior. Finally, realistic standard compliant test signals were utilized for accurate characterization and predistortion of power amplifiers and transmitters.

At this point, it is useful to illustrate a generic block diagram of the experimental setup used for the characterization of power amplifiers' and transmitters' nonlinear behavior. As depicted in Figure 8.2, such a measurement system typically consists of a stimulus generator, a pre-amplifier, the DUT, an output attenuator, and a signal acquisition system. The stimulus generator synthesizes the test signal to be used during the DUT characterization. If needed, a pre-amplifier can be used to adjust the signal level in order to ensure appropriate power levels at the input of the DUT while operating the signal generator in its optimal power range. An attenuator is used at the output of the DUT to condition the power level of the output signal within the dynamic range of the signal acquisition instrument that will be used to collect the output signal. Depending on the type of measurements used, the nature of the instruments used for the stimulus generation and the output signal acquisition can vary.

8.2.1   Characterization Using Continuous Wave Signals

Continuous wave test signals are commonly employed during the design process of power amplifiers. Their use was thus naturally extended to the characterization of

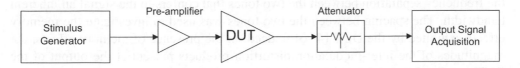

Figure 8.2   Generic block diagram of experimental setup for power amplifiers characterization

power amplifiers and transmitters nonlinearity. A CW signal is fully defined by two parameters: its frequency and power level. Thus, to characterize the power dependent nonlinear behavior of power amplifiers, the experimental approach requires sweeping the power level of the continuous wave test signal. At the output of the DUT, scalar measurements using a spectrum analyzer or a power meter, for example, can be performed to derive the AM/AM (amplitude modulation to amplitude modulation) characteristic of the DUT. In order to capture both the AM/AM and the AM/PM (amplitude modulation to phase modulation) characteristics of the DUT, complex measurements of both the magnitude and phase are needed. Typical continuous wave measurements of power amplifiers' complex gain can be performed using a vector network analyzer.

Even though the experimental setup and the test procedure involved during the characterization of power amplifiers' nonlinear behavior using continuous wave test signals are straightforward, the use of such test signals is undeniably a rudimentary approach that has several limitations. First, this characterization can lead to thermal and bias effects that are mainly stimulated by the nature of the test signal rather than the behavior of the DUT [1]. Furthermore, the continuous wave signal has no inherent bandwidth and accordingly does not provide accurate insight on the DUT dynamic behavior when driven by modulated signals. In fact, for a continuous wave input signal, the nonlinearity of the power amplifier or transmitter causes in-band distortions at the fundamental frequency and out-of-band distortions at the harmonics. The limitation of continuous wave based power amplifiers' characterization for behavioral modeling and predistortion was recognized in [2] back in 1989, where it was reported that limited linearity enhancement on two-tone inter-modulations is obtained when the predistortion function is derived from continuous wave based measurements.

8.2.2 Characterization Using Two-Tone Signals

Due to the deficient nature of the continuous wave signal spectrum, no inter-modulation distortions are present at the output of the nonlinear power amplifier. Thus, a stimulus having a richer spectral content and significant amplitude variation should be applied to excite the nonlinearities and the dynamics of the system, and thus provide a more comprehensive description of the DUT nonlinearity. In this context, the two-tone test signal can be used for the characterization of bandpass nonlinearities as those exhibited by power amplifiers and transmitters [3].

In two-tone test signals, one can control the power level of each tone as well as their relative phase shift. Most importantly, a key aspect of the two-tone stimuli is the frequency separation between the two tones that confers to the signal an inherent bandwidth. The spacing between the two tones was used to investigate the memory effects exhibited by the DUT [4–6]. Indeed, in the presence of memory effects, the magnitudes of the inter-modulation distortion products present at the output of the DUT are found to be dependent on the tone spacing that represents the modulation bandwidth of this test signal. This can be observed experimentally by maintaining a constant power level at the input of the DUT and measuring the third and fifth order inter-modulation distortions levels while varying the tone spacing. Also, asymmetry in the inter-modulation distortion products is observed when memory effects are present in the DUT [7].

However, to maintain the superiority of two-tone based characterization over continuous wave based measurements, it is essential to observe both the AM/AM and AM/PM characteristics of the DUT. This is a major challenge associated with the use of two-tone test signals for power amplifier and transmitter characterization applications. Several experimental setups have been proposed in the literature [4, 8, 9]. These require a reference inter-modulation generator in addition to two signal generators and two spectrum analyzers [8, 9]. However, with the advancement of test and measurement systems, an experimental setup employing an arbitrary waveform generator and a spectrum analyzer having demodulation capabilities was proposed in [4]. The arbitrary waveform generator is applied for accurate generation of amplitude and phase aligned two-tone test signal. The output signal is demodulated using a vector signal analyzer (spectrum analyzer with a built-in modulation analysis feature), and then processed to extract the magnitudes and phases of all relevant frequency components of the output signal. This technique is more accurate than those based on the use of reference inter-modulation generator as it enables the extraction of magnitude and phase information not only for the third order in-band inter-modulation products but also for higher order (fifth and even seventh orders) in-band inter-modulation products.

8.2.3   *Characterization Using Multi-Tone Signals*

The multi-tone test signal is a further enhancement to the two-tone stimulus signal described in the previous section. Indeed, compared to continuous wave and two-tone test signals, multi-tone signals better approximate the frequency content of modern communication signals. The similarity between the behavior of the DUT for a modulated signal excitation and for a multi-tone excitation has been thoroughly investigated in the literature [10–12]. The main conclusion is that special care needs to be taken while engineering multi-tone excitation signals to ensure that they will emulate a behavior of the DUT that is similar to what it would have been for a modulated signal having equal bandwidth and average power.

A multi-tone excitation is determined by the number of tones it contains, their spacing (or equivalently the total bandwidth of the excitation signal), in addition to the relative magnitudes and phases of the tones. The selection of the two first parameters is somehow straightforward. In fact, to approximate a predefined modulated signal, it is obvious that the multi-tone excitation needs to have the same bandwidth. Having defined the bandwidth of the multi-tone excitation, the number of tones can be derived such that the multi-tone excitation has enough resolution in frequency domain. Alternately, a desired tone spacing can be defined and then the required number of tones calculated. The most challenging task in the design of multi-tone excitations for the characterization of power amplifiers' and transmitters' nonlinear behavior is the selection of the tones' magnitudes and phases. In [10], the ability of multi-tone signals with constant and/or random amplitude and phase spectra to generate accurate behavioral models was assessed. The same type of multi-tone signals was used to predict the ACPR (Adjacent Channel Power Ratio) at the output of a power amplifier driven by a modulated signal in [11]. These studies demonstrated that the use of multi-tone test signals can lead to overestimation or underestimation of the spectrum regrowth caused by the DUT when driven by modulated signals, and that it is recommended to generate the multi-tone signal using random distribution for the phases. Guidelines for the optimal design of the amplitudes and phases vectors of multi-tone stimuli that better approximate modulated signals were reported in [12].

Accordingly, carefully designed multi-tone signals can be applied to predict the behavior of nonlinear power amplifiers under modulated test signals. In this type stimulus, the optimization of the tones' phases affects the time domain signal and can significantly change its peak to average power ratio and shape the probability density function of its magnitude to match that of the standard compliant test signal. This will result in good agreement between the responses of the nonlinear DUT to the multi-tone signal and its modulated counterpart.

8.2.4   Characterization Using Modulated Signals

The characterization of power amplifiers' and transmitters' nonlinear behavior using the modulated signal that will be handled by the system in normal mode of operation and during the linearization step is undeniably the most accurate and reliable approach that will lead to satisfactory modeling and linearization performances. The use of the complex baseband waveforms for the characterization of nonlinear amplifiers and transmitters was initially proposed in [13]. The adoption of this approach was made possible thanks to the development of arbitrary waveform generation and vector signal analysis capabilities in test and measurement instruments, and their ability to cope with the bandwidth requirements of communication signals.

The characterization of power amplifiers' and transmitters' nonlinear behavior using modulated signals consists of acquiring the input and output baseband waveforms associated with the bandpass RF (radio frequency) input and output signals of the DUT [14]. This can be implemented according to either of the two schemes depicted in Figure 8.3. In the first scheme illustrated in Figure 8.3a, the input and output RF signals of the power amplifier are acquired and processed to extract the corresponding baseband waveforms that will be used for to identify the behavioral model and/or DPD function. In the second scheme reported in Figure 8.3b, the digital waveform used to build the amplifier's RF input signal is considered as the complex baseband input waveform of the DUT, while the complex baseband output waveform is measured at the output of the DUT. In this case, the reference plane for the input signal measurement is shifted from the input of the power amplifier to the input of the digital to analog

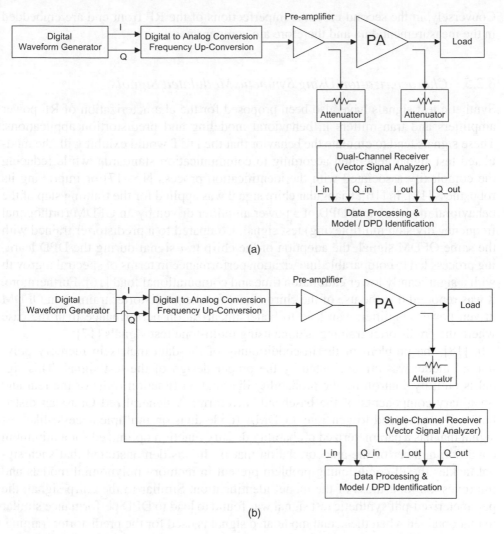

Figure 8.3 Block diagram of experimental setup for nonlinear behavior characterization through complex baseband waveforms measurements. (a) PA only characterization and (b) analog front-end and PA characterization

converter of the signal generator. Thus, the characterization data encompasses the effects of power amplifier as well as all the preceding analog components of the signal generator or equivalently the RF front end. Accordingly, the main difference between the two measurement practices described here is that the first only characterizes the power amplifier, and thus might have limited performance in DPD context when the signal generation path from the output of the digital predistorter up to the input signal measurement reference plane contains some imperfections. Indeed, such imperfections are not included in the observation path and thus cannot be compensated for.

Conversely, in the second case, the imperfections of the RF front end are embedded in the measurement data and therefore are compensated for.

8.2.5 Characterization Using Synthetic Modulated Signals

Synthetic test signals have also been proposed for the characterization of RF power amplifiers and transmitters in behavioral modeling and predistortion applications. These signals tend to emulate the behavior that the DUT would exhibit with the modulated test signal generated according to communication standards, while reducing the complexity associated with the identification process [15–17] or improving its robustness [18]. In [16], triangular chirp signal was applied for the training step of the behavioral modeling and DPD of a power amplifier driven by an OFDM (orthogonal frequency division multiplexing) test signal. Compared to a predistorter trained with the same OFDM signal, the adoption of the chirp test signal during the DPD learning process led to comparable linearization performance in terms of spectral regrowth with a significantly lower calibration time and computational load [16]. Furthermore, it was reported that the use of the chirp signal for the predistorter training in OFDM driven power amplifiers can lead to better linearization performance than the case where the predistorter training is done using multi-tone test signals [17].

In [18], the problem of the ill-conditioning of the data matrix in memory polynomial model was circumvented by the proper design of the test signal. This signal is built by controlling the probability distribution function (pdf) of the real and imaginary components of the baseband waveform. A generalized Gaussian distribution was proposed to generate a CDMA (code division multiple access)-like test signal that has a parameterized probability density function optimized for a minimum condition number of the autocorrelation matrix. It was demonstrated that such signal tackles the ill-conditioning problem present in memory polynomial models and improves the robustness of the model identification. Similar to the chirp signal, the parameterized-pdf synthetic test signal was found to lead to DPD performance similar to that obtained when the actual modulated signal is used for the predistorter training.

8.2.6 Discussion: Impact of Test Signal on the Measured AM/AM and AM/PM Characteristics

The AM/AM and AM/PM characteristics of power amplifiers and transmitters are quite sensitive to the type of drive signal and to its characteristics. In [14], the AM/AM and AM/PM characteristics of a power amplifier prototype were derived for three different excitations signals: a continuous wave, an eight-tone, and a WCDMA (wideband code division multiple access) signal. As illustrated in Figure 8.4, there is significant discrepancy between the characteristics measured with the modulated signal and those measured with the continuous wave signal. However, the multi-tone test signal leads to closer prediction of the DUT nonlinear characteristics. This corroborates the conclusions mentioned in the previous

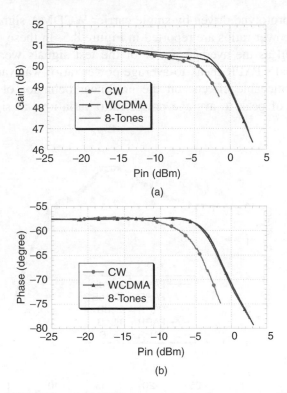

Figure 8.4 Measured AM/AM and AM/PM characteristics of a power amplifier prototype for various excitation signals. (a) AM/AM characteristics and (b) AM/PM characteristics [14]. ©2009 IEEE. Reprinted, with permission, from S. Boumaiza *et al.*, "Systematic and adaptive characterization approach for behavior modeling and correction of dynamic nonlinear transmitters," *IEEE Transactions on Instrumentation and Measurement*, Dec. 2007

sub-sections according to which the use of the modulated signal is more appropriate for accurate characterization of the DUT's nonlinear behavior.

A closer look at the sensitivity of power amplifiers' and transmitters' dynamic nonlinear behavior to the excitation signal reveals that even for a given type of modulated signals, such as CDMA or OFDM, for example, the signal characteristics might noticeably impact the response of the DUT. The characteristics of modulated signals mainly include their average power, bandwidth, and statistics. To describe the statistics of modern communication signals, the complementary cumulative distribution function (CCDF) is widely adopted. The CCDF provides a thorough portrayal of the signal statistics by depicting for each power level above the average power of the signal and the percentage of time the signal power is at or above that value. For a given communication standard, the CCDF curves of all signals are quite similar and variations only occur toward the tail of the curve as the peak to average power ratios of the signals vary. However, these changes are imperceptible by the power amplifier since the CCDF disparity occurs for signal samples with very low probability. The measured memoryless AM/AM and AM/PM characteristics of a power amplifier prototype driven by single carrier WCDMA signals having various peak to average power ratios are reported in Figure 8.5. In these measurements, the bandwidth as well as the average power of the test signals were kept unchanged and only the signal's PAPR (peak-to-average power ratio) was varied. These results confirm the unnoticeable effects, on the nonlinear behavior of power amplifiers and transmitters, of peak to average power ratio variations in signals of the same

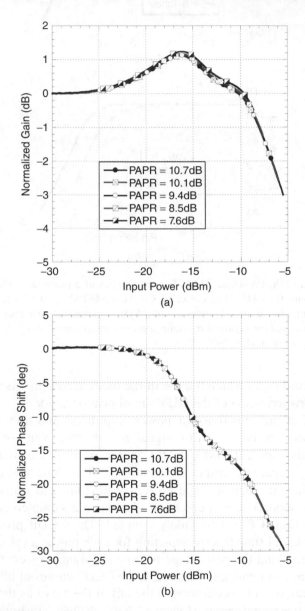

Figure 8.5   Measured AM/AM and AM/PM characteristics of a power amplifier prototype for WCDMA signals with various PAPR. (a) AM/AM characteristics. (b) AM/PM characteristics

standard [19]. However, it is important to mention that any inconsistency in the CCDF characteristics of excitation signals will lead to different nonlinear behaviors of the DUT even if the PAPRs of these signals are comparable. This is the main reason that calls for the careful engineering of multi-tone test signals with particular attention to the distribution of their phases.

Another parameter of the modulated test signal is its average power. The behavior of power amplifiers is sensitive to variations in the average power of the input signal as investigated in [19]. Figure 8.6 depicts the memoryless AM/AM and AM/PM characteristics of a Doherty power amplifier prototype measured for single carrier WCDMA waveforms having different peak to average power ratios and average powers. The average power of each signal was set such that the DUT is operated over its entire power range up to but not beyond its saturation. Accordingly, for each signal, the DUT was driven at an output power back-off that is equal to the signal's peak to average power ratio. The substantial changes in the measured characteristics, especially the AM/AM ones, reveal the dependency of power amplifiers' and transmitters' nonlinear behavior to the average power of the drive signal. This dependency can be solely attributed to the average power variation since no changes in the DUT nonlinearity was observed when the same signals were applied at constant average power (Figure 8.5). Similarly, the nonlinear behavior of power amplifiers and transmitters depend on the signal's bandwidth, which has a considerable impact on the memory effects generated by the DUT. As the drive signal bandwidth increases, the memory effects exhibited by the DUT become more pronounced. This translates into more significant dispersion in the measured AM/AM and AM/PM [20]. Figure 8.7 illustrates an example of the influence of the signal bandwidth on the measured AM/AM characteristics of a PA using LTE (Long-Term Evolution) signals having similar average power but different bandwidths. Similar effect is observed in the AM/PM characteristics.

8.3   Data De-Embedding in Modulated Signal Based Characterization

The raw measured data needs to be processed to extract the corresponding signals at the input and output reference planes of the DUT. This data processing is twofold: power adjustment and time alignment. The power adjustment is straightforward and consists of compensating for the amplifications and mainly attenuations incurred by the signal between the measurement planes and the DUT reference planes.

The time alignment is a critical task in the de-embedding process of the measured data. In fact, the measured output signal is a time delayed nonlinearly amplified version of the input signal. The delay between the measured input and output waveforms is associated with the signal propagation time through the DUT. Figure 8.8 shows the characterization results of a power amplifier driven by an LTE signal before delay compensation. The time domain waveforms of Figure 8.8a clearly reveal the delay between the input and output waveforms. Extracting the AM/AM characteristic of the DUT from these time misaligned waveforms leads to the results reported in Figure 8.8b. This figure illustrates the detrimental effect of the residual delay between

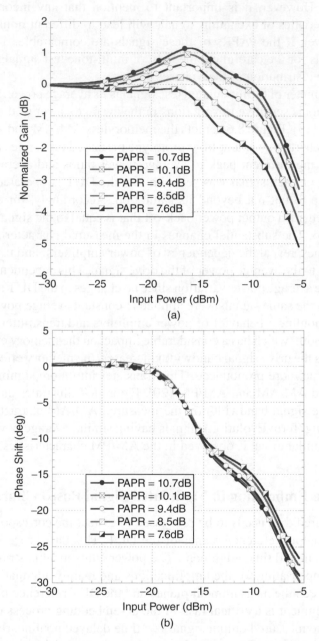

Figure 8.6 Measured AM/AM and AM/PM characteristics of a power amplifier prototype for WCDMA signal with various average power levels. (a) AM/AM characteristics and (b) AM/PM characteristics

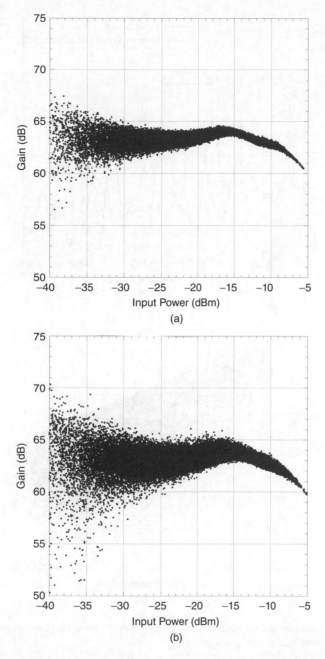

Figure 8.7 Measured AM/AM characteristics of a power amplifier prototype for LTE signals with various bandwidths. (a) One-carrier (20 MHz bandwidth) and (b) three-carrier (60 MHz bandwidth)

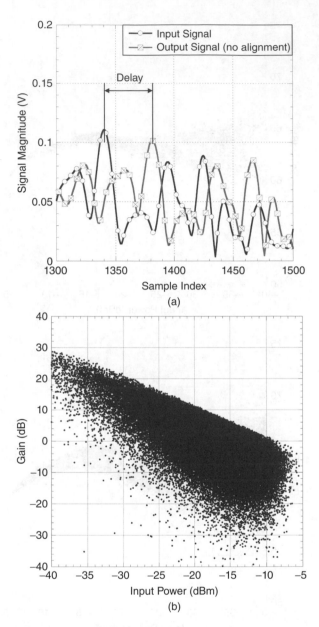

Figure 8.8   PA characterization results before delay compensation. (a) Time domain waveforms and (b) AM/AM characteristic

the input and output waveforms used to generate the AM/AM characteristic of the DUT. Indeed, the time misalignment appears as a dispersion in the AM/AM curve of the DUT. Similar observation can be made for the AM/PM characteristic. This dispersion can be explained by the fact that the delay between the input and output waveforms causes the output sample at index n to heavily depend on the input sample at index $n - D$ where D represents the delay value in terms of the number of samples. Accordingly, time misalignment between the input and output waveforms appears as additional "memory effects." Thus, it manifests in the AM/AM and AM/PM characteristics as a dispersion similar to that observed when memory effects are present.

The widely adopted approach for the estimation and compensation of the propagation delay between the input and output waveforms of the DUT is based on the cross-covariance between these two waveforms [21]. The cross-covariance between the measured input and output waveforms ($x(n)$ and $y(n)$, respectively) is given by:

$$C_{xy}(d) = \begin{cases} \displaystyle\sum_{n=0}^{N-d-1} \left(x(n+d) - \bar{x}\right)\left(y^*(n) - \bar{y}^*\right) & \text{for } d \geq 0 \\ \displaystyle\sum_{n=0}^{N-d-1} \left(x^*(n-d) - \bar{x}^*\right)\left(y(n) - \bar{y}\right) & \text{for } d < 0 \end{cases} \tag{8.1}$$

where N represents the number of data samples, in each of the waveforms, used to estimate the propagation delay. $x^*(n)$ and $y^*(n)$ denote the complex conjugate of the complex waveforms $x(n)$ and $y(n)$, respectively. $\bar{x}$ and $\bar{y}$ are the averages of the waveforms $x(n)$ and $y(n)$, respectively, and are defined according to:

$$\begin{cases} \displaystyle\bar{x} = \frac{1}{N}\sum_{n=0}^{N-1} x(n) \\ \displaystyle\bar{y} = \frac{1}{N}\sum_{n=0}^{N-1} y(n) \end{cases} \tag{8.2}$$

Equation 8.1 estimates the delay between the input and output waveforms in terms of number of samples. The delay corresponds to the value of d for which C_{xy} is maximum. However, given the typical sampling rates used for the acquisition of these waveforms, the resulting delay resolution is usually in the range of several nanoseconds. In fact, a sampling rate of 100 MHz corresponds to a delay resolution of 10 ns. Thus, performing the delay alignment process using the original sampling frequency of the measured waveforms is generally referred to as coarse delay alignment as it results in a substantial residual delay. The data of Figure 8.8 was acquired at a sampling rate of 245.76 MHz that corresponds to delay resolution in the range of 4.08 ns. The input and output waveforms obtained after performing the coarse delay alignment technique are presented in Figure 8.9a, and the AM/AM characteristic derived from these aligned waveforms is reported in Figure 8.9b. Compared to what was observed in Figure 8.8, it is clear that the delay alignment allows for the elimination of most of the dispersion present in the AM/AM characteristic derived from the raw measured data.

In [21], it was revealed that a sub-sample resolution is needed during the delay estimation and compensation process in order to fully eliminate the effects of the time misalignment between the measured input and output waveforms. This is performed by applying the delay estimation technique described by Equation 8.1 on the oversampled version of the input and output waveforms. Accordingly, the raw

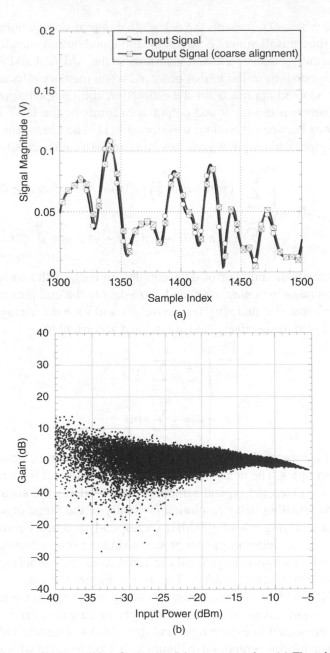

Figure 8.9 PA characterization results after coarse delay compensation. (a) Time domain waveforms and (b) AM/AM characteristic

data used to derive the results of Figure 8.8 were oversampled with a ratio of 25, and then the delay estimation and compensation was performed using the oversampled waveforms. The results are summarized in Figure 8.10 showing both the time domain waveforms and the corresponding AM/AM characteristic of the DUT. This

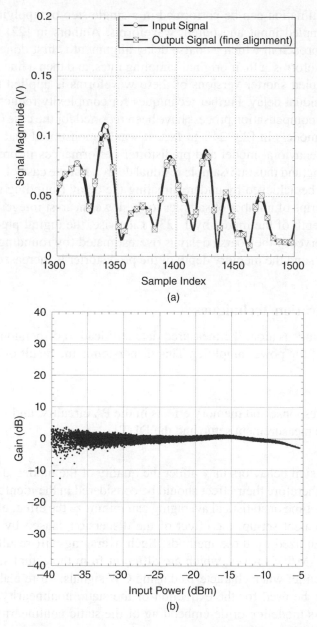

Figure 8.10 PA characterization results after fine delay compensation. (a) Time domain waveforms and (b) AM/AM characteristic

figure demonstrates the effectiveness of the sub-sample delay resolution as further noticeable reduction in the dispersion of the AM/AM characteristic is obtained in comparison with the results of Figure 8.9. This justifies the usefulness of this approach often labeled fine delay alignment and regularly used in power amplifier behavioral modeling and predistortion applications.

The delay estimation can be performed in a single step by applying Equation 8.1 on the oversampled input and output waveforms. Authors in [22] proposed a low complexity approach in which a coarse delay alignment is first done using the input and output waveforms at their original sampling rates, and then a fine delay alignment using oversampled shorter versions of these waveforms is applied to eliminate the sub-sample residual delay. Further techniques for complexity reduction in the delay estimation and compensation process have been reported for the case of memory polynomial based models in [23, 24]. In fact, it was demonstrated that it is possible to maintain the behavioral model and predistorter performances in presence of coarse delay alignment (and thus sub-sample residual delay). For the case of behavioral modeling, this can be achieved by underestimating the delay (rounding the coarse delay expressed in terms of number of samples to the lower nearest integer) and increasing the memory depth of the model by 1 [23]. Likewise, the digital predistorter performance is preserved if the coarse delay is overestimated (by rounding it to the higher nearest integer) and the memory depth of the predistorter is increased by 1 [24].

8.4   Identification Techniques

After delay compensation, the measured data still lead to dispersion in the AM/AM and AM/PM of the power amplifier. This dispersion is the result of the contribution of:

- Frequency response and memory effects in the PA circuitry, and
- Noise in the measurement setup and the DUT.

These dispersion behaviors may affect the quality of the model identification considerably and therefore their effect should be considered in the identification process. While a simple time or statistical averaging can minimize the effect of the white noise in the measurement set-up, the effect of the dispersion caused by memory effects cannot be minimized by these methods. Such averaging will result in non-smooth AM/AM and AM/PM curves, which are different from the actual static response of the power amplifier when characterized using CW signals. More elaborate averaging techniques can be used for the extraction of the static nonlinearity. This is needed for memoryless modeling or de-embedding of the static nonlinearity from dynamic nonlinearities for box-oriented modeling.

8.4.1   Moving Average Techniques

In the first step, the AM/AM and AM/PM of the power amplifier or nonlinear transmitter are plotted using the measured data after compensating for the delay in the transmission and feedback paths. These plots include significant dispersion caused by the dynamic response of the system (frequency response and memory effects) along with the effect of noise generated by the DUT and the measurement set up.

The next step consists of isolating the static nonlinear behavior from the dynamic effect causing the dispersion. Such operation is achieved using averaging. This operation consists of reducing the dispersion by averaging the gain and phase values for the same input power. This statistical averaging will eliminate the dispersion caused by noise and will remove the dispersion caused by memory effects. It consists first of sorting all the samples of the measured AM/AM and AM/PM data according to the power of the input signal. Then, a statistical average is applied for the data with the same input power. Each of the obtained AM/AM and AM/PM curves becomes a single line with no dispersion in it. However, this line is not smooth and does not represent accurately the static nonlinearity of the system. It includes residual dispersive behavior that is averaged statistically. If such curve were to be used to de-embed the static nonlinearity from the dynamic response (frequency response) for box oriented models, modeling will not be accurate.

In a final step, a moving average algorithm can be used to remove the residual dispersive behavior from the statically averaged signal.

Different moving average algorithms were proposed in the literature to smooth the AM/AM and AM/PM characteristics of power amplifiers. In [21, 25, 26], Liu *et al.* propose to use the following algorithm for the moving average:

$$\hat{g}(n) = \alpha \cdot g(n) + (1 - \alpha)\hat{g}(n - 1) \tag{8.3}$$

where $g(n)$ is the power-sorted and time-averaged gain at the input of the moving average block, $\hat{g}(n)$ is the output of the moving average algorithm, n refers to the index of the gain value in the power-sorted gain vector, and α is a weighting factor that should take a value between 0 and 1. In the simplest situation, α can be constant. However, Liu *et al.* stated that given the significant dispersion in the AM/AM and AM/PM characteristics of wideband power amplifiers, a fixed weighting factor along the full dynamic range of the signal often leads to a poor moving average quality. Indeed, if α is large (close to 1), the output of the moving average algorithm is very close to its input; a very small amount of averaging is applied to the signal and the output is not properly smoothed. Contrarily, if α is small (close to 0), the output will be smoothed but may result in an incorrect traces due to the average error propagation. Liu *et al.* proposed to use an exponentially weighted moving average, in which the value of α is changing to adapt to the changes in the AM/AM and AM/PM characteristics. More precisely, in the proposed weighted moving average algorithm, α is a function of the input power, that is, $\alpha = F[|x^2(n)|]$.

In [27, 28], authors propose to use a more dynamic moving average algorithm. This moving average algorithm can be represented by the following equations:

$$\tilde{g}(n) = \hat{g}(n - 1) + \frac{x(n) - x(n - 1)}{x(n + 1) - x(n - 1)} \cdot [g(n + 1) - \hat{g}(n - 1)] \tag{8.4}$$

$$\hat{g}(n) = \lambda(n) \cdot g(n) + [1 - \lambda(n)]\tilde{g}(n) \tag{8.5}$$

In this algorithm, the averaging is done in two steps. The first step consists of a static moving average where the resulting signal $\tilde{g}(n)$ is an average between the future sample of the input signal, $g(n+1)$, and the past sample of the moving average block output, $\hat{g}(n-1)$. The constant weighting factor is only a function of the step size of power, making the resulting signal $\tilde{g}(n)$ a linear interpolation between $g(n+1)$ and $\hat{g}(n-1)$. The second step consists of a dynamic moving average where the output signal $\hat{g}(n)$ is provided by a dynamically weighted average between $\tilde{g}(n)$ and the present sample of the input signal $g(n)$. $\lambda(n)$ is the regression factor, which can take values between 0 and 1. For values close to 1, small changes are applied to the output signal and therefore non smoothed curve is obtained. Kwan *et al.* [28, 29] showed that choosing values of $\lambda(n)$ proportional to the second derivative of the gain function (AM/AM or AM/PM) leads often to better averaging performance.

8.4.2   Model Coefficient Extraction Techniques

In Chapters 4, 5, and 6, the different power amplifier models with and without memory were introduced and formulated mathematically. Depending on their mathematical formulation, each block of these models can be classified in either of the two following categories:

- Look-up table based blocks, where the nonlinear function is implemented using look-up tables. These look-up tables are built using measured data in order to include information on the nonlinear behavior of the block. In this case, no coefficients need to be extracted for modeling. The model identification consists of building the look-up table content, which is obtained from processing the measured data as shown in the previous sections of this chapter. This operation includes the delay compensation, data de-embedding, and moving average techniques.
- Equation based blocks, where the block is modeled by an equation relating the output signal of the block to its input signal. Such an equation can model nonlinearity (e.g., memoryless polynomial), memory effects (e.g., finite impulse response filter in Weiner model), or the joint effects of memory and nonlinearity (e.g., memory polynomial). This equation has a certain form and includes coefficients that are device dependent. In this case, the model identification consists of extracting these coefficients.

In this section, we will present the general formulation for an equation based model and discuss the most frequently used algorithms for coefficient extraction.

If we consider the memory polynomial model, the output is represented as a function of the input signal using the equation below as was shown in Equation 5.2:

$$y(n) = \sum_{m=0}^{M} \sum_{k=1}^{K} a_{mk} \cdot x(n-m) \cdot |x(n-m)|^{k-1} \tag{8.6}$$

where $y(n)$ is the model's baseband complex output sample at instant n and $x(n - m)$ is the model's baseband complex input sample at instant $n - m$. It was shown in Chapter 5 that this representation can be rewritten in vector format as:

$$y(n) = \boldsymbol{\varphi}_{MP}(n) \cdot \mathbf{A} \tag{8.7}$$

where $\boldsymbol{\varphi}_{MP}(n)$ is a vector built using the baseband complex input signal samples $x(n - m)$ according to the model's basis functions set, and $\mathbf{A}$ is the vector containing the model coefficients. These are given by:

$$\boldsymbol{\varphi}_{MP}(n) = [x(n) \cdots x(n) \cdot |x(n)|^{K-1} x(n-1) \cdots x(n-1)\cdot$$
$$|x(n-1)|^{K-1} \cdots x(n-M) \cdot |x(n-M)|^{K-1}] \tag{8.8}$$

$$\mathbf{A} = \begin{bmatrix} a_{01} & \cdots & a_{0K} & a_{11} & \cdots & a_{1K} & \cdots & a_{MK} \end{bmatrix}^T \tag{8.9}$$

where $[]^T$ denotes the transpose operator.

For a set of N samples, this vector representation can then be rewritten in matrix format as follows:

$$\mathbf{y} = \mathbf{X} \cdot \mathbf{A} \tag{8.10}$$

where $\mathbf{y} = [y(n)\, y(n-1) \cdots y(n-N+1)]^T$ is the vector of N samples of the output signal, and $\mathbf{X}$ is a matrix whose rows are delayed versions of $\boldsymbol{\varphi}_{MP}(n)$. It is given by:

$$\mathbf{X} = [\boldsymbol{\varphi}_{MP}(n)\, \boldsymbol{\varphi}_{MP}(n-1) \cdots \boldsymbol{\varphi}_{MP}(n-N+1)]^T$$

$$= \begin{bmatrix} x(n) & \cdots & x(n) \cdot |x(n)|^{K-1} & x(n-1) & \cdots & x(n-1) \cdot |x(n-1)|^{K-1} \\ x(n-1) & \cdots & x(n-1) \cdot |x(n-1)|^{K-1} & x(n-2) & \cdots & x(n-2) \cdot |x(n-2)|^{K-1} \\ \vdots & \cdots & \vdots & \vdots & \cdots & \vdots \\ x(n-N+1) & \cdots & x(n-N+1) \cdot |x(n-N+1)|^{K-1} & x(n-N) & \cdots & x(n-N) \cdot |x(n-N)|^{K-1} \end{bmatrix}$$

$$\begin{bmatrix} \cdots & x(n-M) \cdot |x(n-M)|^{K-1} \\ \cdots & x(n-M-1) \cdot |x(n-M-1)|^{K-1} \\ \cdots & \vdots \\ \cdots & x(n-M-N+1) \cdot |x(n-M-N+1)|^{K-1} \end{bmatrix} \tag{8.11}$$

This matrix formulation can be carried out for any polynomial based model or block of a multi-box model. The only change that will occur from one model to another will be in the matrix composition of the data matrix, $\mathbf{X}$. Given the matrix formulation of the model, the coefficients identification corresponds to calculating the vector $\mathbf{A}$. If the matrix $\mathbf{X}$ was invertible, the coefficients identification would be given by:

$$\mathbf{A} = \mathbf{X}^{-1} \cdot \mathbf{y}. \tag{8.12}$$

However, this is not the case, and the system provided by Equation 8.10 corresponds to an over-determined system. An approximate solution to the coefficients extraction can be obtained by minimizing the mean squared error, e, given by:

$$e = \|\mathbf{y} - \mathbf{X}\mathbf{A}\|^2. \tag{8.13}$$

One approach that is commonly used to solve for this problem consists of computing the Moore−Penrose pseudo-inverse of the matrix $\mathbf{X}$ given by [30−32]:

$$pinv(\mathbf{X}) = (\mathbf{X}^T\mathbf{X})^{-1}\mathbf{X}^T. \tag{8.14}$$

The coefficients will then be calculated using:

$$\mathbf{A} = pinv(\mathbf{X}) \cdot \mathbf{y}. \tag{8.15}$$

This technique allows for a good estimation of the model coefficients by minimizing the error using least squares criterion. It also offers relatively good stability and convergence performance. However, the main drawback is the computational complexity. Indeed, it requires a matrix inversion along with matrices multiplications. Even though singular value decomposition (SVD) is used to facilitate the matrix inversion calculation, the resulting complexity prevent the use of this technique in continuously adaptive and online linearization.

In order to reduce the computational complexity, adaptive filtering algorithms can be used to replace the SVD algorithm. One can classify these algorithms into two different classes:

1. The stochastic gradient family, which includes the least-mean-squares (LMSs) algorithm and its variations such as the normalized LMS algorithm, the leaky LMS algorithm, and so on.
2. The recursive least-squares (RLSs) family, which includes the RLSs algorithm and its variations such as the QR decomposition based recursive least-square (QR-RLS) and the exponentially-weighted RLS.

The concept of adaptive filtering consists of calculating iteratively the model output using the instantaneous error and with an objective of minimizing the power of this error, or in other words, minimizing the mean squared error. For PA behavioral modeling, the process is conceptually a system identification problem where the model intends to replicate the system (PA) behavior. In this case, the system and the model have the same input $x(n)$ and the adaptive filter output is an estimate of the model output, $\hat{y}(n)$. The desired output of the filter $(d(n))$ is therefore the output of the system:

$$d(n) = y(n) \tag{8.16}$$

and the modeling error is therefore the difference between the output of the adaptive filter and the output of the system:

$$e(n) = d(n) - \hat{y}(n) = d(n) - \boldsymbol{\varphi}_x(n) \cdot \mathbf{A}(n) \tag{8.17}$$

where $\boldsymbol{\varphi}_x(n)$ is a vector built using the baseband complex input signal samples $x(n - m)$ according to the model's basis functions set. For example, in the case of memory polynomial, this vector is $\boldsymbol{\varphi}_x(n)$ is given by:

$$\boldsymbol{\varphi}_x(n) = \boldsymbol{\varphi}_{MP}(n) = [x(n) \cdots x(n) \cdot |x(n)|^{K-1} \, x(n-1) \cdots$$

$$x(n-1) \cdot |x(n-1)|^{K-1} \cdots x(n-M) \cdot |x(n-M)|^{K-1}] \quad (8.18)$$

and $A(n)$ is the vector containing the adaptive filter coefficients at instant n. A block diagram of the concept of system identification is presented in Figure 8.11.

In the case of DPD, the problem is cast as a reverse modeling of power amplifiers. In this case, the post-inverse is identified and used for predistortion purpose. In this case, the gain normalized output of the system (PA), $y(n)/G$, where G is the linear gain of the PA is used as the input of the adaptive filter. The adaptive filter provides at its output an estimate, $\hat{y}(n)$, of the input signal to the PA, $x(n)$. The desired output of the filter is therefore the delayed version of the input of the system:

$$d(n) = x(n-\beta) \quad (8.19)$$

and the modeling error is therefore the difference between the output of the adaptive filter and the input of the system:

$$e(n) = d(n) - \hat{y}(n) = d(n) - \boldsymbol{\varphi}_y(n) \cdot A(n) \quad (8.20)$$

where $\boldsymbol{\varphi}_y(n)$ is a vector built using the baseband complex output signal samples of the system, $y(n-m)$, according to the model's basis functions set. For example, in the case of memory polynomial, this vector is given by:

$$\boldsymbol{\varphi}_y(n) = \boldsymbol{\varphi}_{MP}(n) = [y(n) \cdots y(n) \cdot |y(n)|^{K-1} \, y(n-1) \cdots y(n-1) \cdot |y(n-1)|^{K-1} \cdots$$

$$y(n-M) \cdot |y(n-M)|^{K-1}] \quad (8.21)$$

and $A(n)$ is the vector containing the adaptive filter coefficients at instant n. A block diagram of the concept of system identification is presented in Figure 8.12.

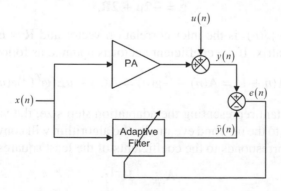

Figure 8.11   Block diagram of the system identification concept in adaptive filtering

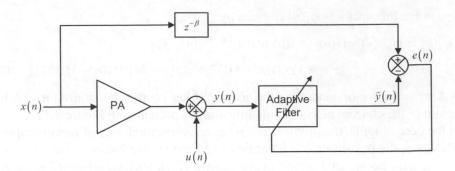

Figure 8.12 Block diagram of the system reverse modeling concept in adaptive filtering

In general, one can formulate the problems of model identification and reverse modeling using the general equation of the error:

$$e(n) = d(n) - \hat{y}(n) = d(n) - \boldsymbol{\varphi}(n) \cdot \mathbf{A}(n) \tag{8.22}$$

where $\boldsymbol{\varphi}(n)$ can be $\boldsymbol{\varphi}_x(n)$ or $\boldsymbol{\varphi}_y(n)$, according to the case, whether it is system identification or reverse modeling, respectively. The objective here is to minimize the mean squared error with regards to $\mathbf{A}(n)$:

$$J[\mathbf{A}(n)] = E[e^2(n)] \tag{8.23}$$

where $E[x]$ is the expectation of x.

Using the steepest descent algorithm, the gradient of this quantity can be shown to be equal to:

$$g = \frac{\partial J(\mathbf{A})}{d\mathbf{A}} = -2E[\boldsymbol{\varphi}^T(n)e(n)] = -2E[\boldsymbol{\varphi}^T(n)d(n) - \boldsymbol{\varphi}^T(n)\boldsymbol{\varphi}(n)\mathbf{A}] \tag{8.24}$$

which can be written as:

$$g = -2\mathbf{p} + 2\mathbf{RA} \tag{8.25}$$

where $\mathbf{p} = E[\boldsymbol{\varphi}^T(n)d(n)]$ is the inter-correlation vector and $\mathbf{R} = E[\boldsymbol{\varphi}^T(n)\boldsymbol{\varphi}(n)]$ is the autocorrelation matrix. If the coefficient vector is updated as follows:

$$\mathbf{A}(n+1) = \mathbf{A}(n) - \frac{\mu}{2}g(n) = \mathbf{A}(n) + \mu E[\boldsymbol{\varphi}^T(n)e(n)] \tag{8.26}$$

where μ is a constant representing the adaptation step size, the value of J decreases from one iteration to the next and eventually the algorithm will converge to the optimal solution, which corresponds to the coefficients of the least squares solution:

$$\mathbf{A}_{opt} = \mathbf{R}^{-1}\mathbf{p} \tag{8.27}$$

In practice, the values of $\mathbf{p}$ and $\mathbf{R}$ cannot be exactly determined, neither can be the expectation $E[\boldsymbol{\varphi}^T(n)e(n)]$. In the stochastic gradient family and the recursive least square family, these deterministic quantities are approximated by estimates.

In the following, an example of use of the adaptive filtering algorithms, LMS and RLS, in the modeling or reverse modeling of power amplifiers will be discussed.

8.4.2.1   The LMS Algorithm

The concept of the LMS algorithm consists of getting the simplest estimates $\hat{R}$ and $\hat{p}$ for the autocorrelation matrix R and the intercorrelation vector p, respectively, which are the instantaneous estimates given by:

$$\hat{R}(n) = \boldsymbol{\varphi}^T(n)\boldsymbol{\varphi}(n) \tag{8.28}$$

$$\hat{p}(n) = \boldsymbol{\varphi}^T(n)d(n) \tag{8.29}$$

The different steps of the algorithm can then be summarized in the following equations:

- *Initialization steps:*

$$A(0) = 0 \tag{8.30}$$

- *nth iteration:*

$$\hat{y}(n) = \boldsymbol{\varphi}(n) \cdot A(n) \tag{8.31}$$

$$e(n) = d(n) - \hat{y}(n) \tag{8.32}$$

$$A(n+1) = A(n) + \mu\boldsymbol{\varphi}^T(n)e(n) \tag{8.33}$$

where μ is a constant that represents the adaptation step size. In practice, this constant has to be carefully chosen in order to guarantee the convergence of the algorithm.

8.4.2.2   The RLS Algorithm

The concept of the RLS algorithm consists of estimating the adaptive filter coefficients in order to minimize the least squares criteria given by:

$$J[A(n)] = \sum_{i=0}^{n} \lambda^{n-i}[d(i) - \boldsymbol{\varphi}(i)A(n)] \tag{8.34}$$

where λ is constant factor that represents the degree of dependence from previous iterations $(0 \ll \lambda < 1)$. It is often called the forgetting factor. In this case, the estimates for the autocorrelation matrix R and the intercorrelation vector p are more realistic for large values of n. They are given by:

$$\hat{p}(n) = \sum_{i=0}^{n} \lambda^{n-i}\boldsymbol{\varphi}^T(i)d(i) \approx p(n) \tag{8.35}$$

$$\hat{\mathbf{R}}(n) = \sum_{i=0}^{n} \lambda^{n-i} \boldsymbol{\varphi}^T(i) \boldsymbol{\varphi}(i) \approx \mathbf{R}(n) \tag{8.36}$$

From the previous two equations, one can write:

$$\mathbf{R}(n) = \lambda \mathbf{R}(n-1) + \boldsymbol{\varphi}^T(n) \boldsymbol{\varphi}(n) \tag{8.37}$$

$$\mathbf{p}(n) = \lambda \mathbf{p}(n-1) + \boldsymbol{\varphi}^T(n) d(n) \tag{8.38}$$

If we note the inverse of the autocorrelation matrix by:

$$\mathbf{Q}(n) = \mathbf{R}^{-1}(n) \tag{8.39}$$

Then, it can be shown that the inverse of the autocorrelation matrix can be calculated recursively as follows:

$$\mathbf{Q}(n) = \lambda^{-1} \mathbf{Q}(n-1) - \frac{\lambda^{-2} \mathbf{Q}(n-1) \boldsymbol{\varphi}^T(n) \boldsymbol{\varphi}(n) \mathbf{Q}(n-1)}{1 + \lambda^{-1} \boldsymbol{\varphi}(n) \mathbf{Q}(n-1) \boldsymbol{\varphi}^T(n)} \tag{8.40}$$

With the estimates of Equations 8.35 and 8.36, the algorithm should converge very close to the optimal solution, which corresponds to the coefficients of the least squares solution:

$$\mathbf{A}_{opt} = \mathbf{R}^{-1} \mathbf{p} \tag{8.41}$$

The different steps of the algorithm can be summarized in the following equations:

- *Initialization steps:*

$$\mathbf{A}(0) = \mathbf{0} \tag{8.42}$$

$$\mathbf{Q}(0) = \delta^{-1} \mathbf{I} \tag{8.43}$$

where $\delta > 0$ is an initialization constant and $\mathbf{I}$ is the identity matrix.
- *nth iteration:*

$$\mathbf{k}(n) = \frac{\lambda^{-1} \mathbf{Q}(n-1) \boldsymbol{\varphi}^T(n)}{1 + \lambda^{-1} \boldsymbol{\varphi}(n) \mathbf{Q}(n-1) \boldsymbol{\varphi}^T(n)} \tag{8.44}$$

$$e(n) = d(n) - \boldsymbol{\varphi}^T(n) \mathbf{h}(n-1) \tag{8.45}$$

$$\mathbf{h}(n) = \mathbf{h}(n-1) + \mathbf{k}(n) e^*(n) \tag{8.46}$$

$$\mathbf{Q}(n) = \lambda^{-1} \mathbf{Q}(n-1) - \lambda^{-1} \mathbf{k}(n) \boldsymbol{\varphi}(n) \mathbf{Q}(n-1) \tag{8.47}$$

where $0 \ll \lambda < 1$ and λ is the forgetting factor.

In general, the stochastic gradient family has lower computational complexity but poorer residual error performance than the recursive least squares family and therefore is less used in nonlinear systems modeling. More detailed information about these algorithms can be obtained in references on adaptive filtering [33–35].

8.5 Robustness of System Identification Algorithms

Each of the previously described identification algorithms is dependent on some criteria to converge. A careful choice of the parameters is necessary to ensure convergence of the algorithm. In the following, a description of the conditions for convergence for each of these algorithms will be provided along with a comparison through an example of robustness, expected performance in terms of amount of residual error, convergence time, and computation complexity for these algorithms.

8.5.1 The LS Algorithm

Given the least-squares (LS) algorithm is a non-iterative algorithm, it has no need for parameter initialization. Therefore, its convergence is independent from such step. However, this algorithm involves the inversion of the autocorrelation matrix $\mathbf{X}^T\mathbf{X}$, which has a size of $L \times L$, where L is the number of the model coefficients. The number of multiplications required for this inversion is in the order of $O(L^3)$.

The computational complexity is not the in only challenge using the LS algorithm. The stability of implementation is also a major challenge. Indeed, it was shown that the autocorrelation matrix $\mathbf{X}^T\mathbf{X}$ is ill-conditioned resulting in significantly large ratio between the highest and lowest eigenvalues. As a result, this matrix is almost singular and its inverse cannot be computed with enough accuracy [36, 37]. It is important to mention that this ill-conditioning problem is intensified with large nonlinear orders of the model. Therefore, the modeling accuracy cannot keep improving indefinitely as the nonlinearity order of the model increases. A tradeoff between the matrix conditioning and the number of coefficients for the model has to be considered in order to achieve the lowest residual errors.

Finally, the LS algorithm, when not faced with an ill-conditioning problem, is expected to provide the optimal solution in the terms of least squares criterion without any additional residual errors and therefore is expected to result in the lowest possible J for a given model.

$$J_{LS} \approx J_{\min} \tag{8.48}$$

8.5.2 The LMS Algorithm

In the LMS algorithm described in the previous section, the parameter μ is a constant that represents the adaptation step size. For small values of μ, the convergence is slow and several iterations are needed for the algorithm to converge. If μ increases, the convergence becomes faster but the risk of divergence can increase. It has been shown that a practical value of μ that guarantees convergence of the LMS algorithm has to satisfy:

$$0 < \mu < \frac{2}{L\sigma^2} \tag{8.49}$$

where L is the number of the adaptive filter coefficients and σ^2 is the variance of the input signal to the adaptive filter.

The LMS algorithm is very simple. It requires only $2L + 1$ multiplications and $2L$ additions per iteration. However, it requires a larger number of iterations than the RLS algorithm to converge and often results in higher residual errors. In fact, the residual error in LMS algorithm can be shown to be equal to [33]:

$$J_{LMS} \approx J_{min} \cdot \left(1 + \frac{\mu}{2} L\sigma^2 \right) \tag{8.50}$$

From Equation 8.50, one can deduce that the LMS algorithm performance depends on the signal statistics.

8.5.3   The RLS Algorithm

In the RLS algorithm described in the previous section, the parameter λ is a constant that represents the forgetting factor or the degree of dependence from previous iterations or memory of the algorithm. If λ gets very close to 1, the memory becomes large, which will enhance its performance in terms of residual error and make it get closer to that of the LS algorithm. However, this will result in slow convergence. Decreasing the value of λ makes the convergence faster but reduces the accuracy of the algorithm. It was shown that, in order to achieve accurate and relatively fast convergence of the RLS algorithm with minimal residual error, it is practically preferable to select λ such that:

$$\lambda > 1 - \frac{1}{3L} \tag{8.51}$$

where L is the number of the adaptive filter coefficients.

The RLS algorithm can therefore result in estimates of the model coefficients that have accuracy very close to that of the LS algorithm while avoiding the computational complexity associated with the matrix inversion needed in the LS algorithm. The number of multiplications required by the RLS algorithms is in the order of $O(L^2)$, which is higher than the computation complexity of the LMS algorithm but significantly lower than the computation complexity in the LS algorithm. The residual error in RLS algorithm can be shown to be equal to [33]:

$$J_{RLS} \approx J_{min} \cdot \left(1 + \frac{1 - \lambda}{2} L \right) \tag{8.52}$$

From the previous equation, it is clear that the residual error in RLS can be very close to the residual error in LS algorithm if the forgetting factor gets very close to 1. More importantly, contrary to residual error in LMS, this residual error is independent from the statistical characteristics of the signal itself.

In the following, an example of applying the three coefficient identification algorithms, LS, RLS, and LMS, for identifying a memory polynomial model is provided to compare these algorithms in terms of normalized mean squared error and time to convergence. In this example, the memory polynomial model was used for reverse

Table 8.1 Performance comparison of LS, RLS, and LMS algorithms

	LS	LMS	RLS
Residual normalized mean squared error (dB)	−42.5	−31.2	−41.7
Convergence time (iterations)	−	350	200

modeling a power amplifier (DPD synthesis). The model has a nonlinearity order equal to 12 and memory depth equal to 3. The comparison is summarized in Table 8.1 [38]. The poor performance of LMS algorithm can be attributed to two factors: (i) the high nonlinearity order used for the model which is 12 results in a very high dispersion in the values of the model coefficients; (ii) the correlated nature of signal's samples.

8.6 Conclusions

In this chapter, practical aspects for power amplifier modeling, including characterization and identification techniques, were presented. It was shown that the power amplifier behavior is significantly affected by the signal statistical characteristics. Therefore, for better model accuracy, it is preferable to characterize the power amplifier using a modulated signal with similar characteristics to the model that will be used for the power amplifier. While such a characterization method achieves better modeling accuracy than using a continuous wave or multi-tone characterization, the use of a modulated signal in the characterization process requires involving more complicated identification techniques compared to a continuous wave characterization.

Two different classes of identification techniques were presented. The first class consists of moving average algorithms used to separate the static nonlinearity from the dynamic behavior of the power amplifier. These algorithms are used to identify memoryless models including look-up table based models or equation based models. The moving average algorithms are also used in the identification of box oriented models that include static nonlinear box(es) within the model.

The second class of model identification is used for the identification of equation based models including the family of memory polynomial based models and their variations. Three different algorithms for the identification of the model coefficients were presented and compared in terms of quality of estimation quantified with the normalized mean squared of the residual error, the robustness of convergence, the computational complexity, and the speed of convergence.

References

[1] Boumaiza, S. and Ghannouchi, F.M. (2002) Realistic power-amplifiers characterization with application to baseband digital predistortion for 3G base stations. *IEEE Transactions on Microwave Theory and Techniques*, **50** (12), 3016–3021.
[2] Bosch, W. and Gatti, G. (1989) Measurement and simulation of memory effects in predistortion linearizers. *IEEE Transactions on Microwave Theory and Techniques*, **37** (12), 1885–1890.

[3] Maseng, T. (1978) On the characterization of a bandpass nonlinearity by two-tone measurements. *IEEE Transactions on Communications*, **26** (6), 746–754.

[4] C. Crespo-Cadenas, J. Reina-Tosina, and M. J. Madero-Ayora, Phase characterization of two-tone intermodulation distortion. Digest 2005 IEEE MTT-S International Microwave Symposium (IMS), Long Beach, CA, June 2005, pp. 1505–1508, 2005.

[5] H. Ku, M. D. McKinley, and J. S. Kenney, Extraction of accurate behavioral models for power amplifiers with memory effects using two-tone measurements. Digest 2002 IEEE MTT-S International Microwave Symposium (IMS), Seattle, WA, June 2002, pp. 139–142, 2002.

[6] Ku, H., McKinley, M.D. and Kenney, J.S. (2002) Quantifying memory effects in RF power amplifiers. *IEEE Transactions on Microwave Theory and Techniques*, **50** (12), 2843–2849.

[7] Carvalho, N.B. and Pedro, J.C. (2002) A comprehensive explanation of distortion sideband asymmetries. *IEEE Transactions on Microwave Theory and Techniques*, **50** (9), 2090–2101.

[8] Suematsu, N., Iyama, Y. and Ishida, O. (1997) Transfer characteristic of IM_3 relative phase for a GaAs FET amplifier. *IEEE Transactions on Microwave Theory and Techniques*, **45** (12), 2509–2514.

[9] Yang, Y., Yi, J., Nam, J. *et al.* (2001) Measurement of two-tone transfer characteristics of high-power amplifiers. *IEEE Transactions on Microwave Theory and Techniques*, **49** (3), 568–571.

[10] D. Schreurs, M. Myslinski, and K. A. Remley, RF behavioural modelling from multisine measurements: influence of excitation type. *Proceedings 2003 European Microwave Conference (EuMC), Munich, Germany*, October 2003, pp. 1011–1014, 2002.

[11] K. A. Remley, Multisine excitation for ACPR measurements. Digest 2003 IEEE MTT-S International Microwave Symposium (IMS), Philadelphia, PA, June 2003, pp. 2141–2144, 2002.

[12] Pedro, J.C. and Carvalho, N.B. (2005) Designing multisine excitations for nonlinear model testing. *IEEE Transactions on Microwave Theory and Techniques*, **53** (1), 45–54.

[13] E. G. Jeckeln, F. Beauregard, M. A. Sawan, and F. M. Ghannouchi, Adaptive baseband/RF predistorter for power amplifiers through instantaneous AM-AM and AM-PM characterization using digital receivers. Digest 2000 IEEE MTT-S International Microwave Symposium (IMS), Boston, MA, June 2000, pp. 489–492, 2000.

[14] Boumaiza, S., Helaoui, M., Hammi, O. *et al.* (2007) Systematic and adaptive characterization approach for behavior modeling and correction of dynamic nonlinear transmitters. *IEEE Transactions on Instrumentation and Measurement*, **56** (6), 2203–2211.

[15] C. Sanchez, J. Mingo, P. Garcia *et al.* Memory behavioral modeling of RF power amplifiers. Proceedings 2008 IEEE Vehicular Technology Conference Spring (VTC-Spring), Singapore, May 2008, pp. 1954–1958, 2008.

[16] L. Aladren, P. Garcia-Ducar, J. Mingo *et al.* Behavioral power amplifier modeling and digital predistorter design with a chirp excitation signal. Proceedings 2011 IEEE Vehicular Technology Conference Spring (VTC-Spring), Budapest, Hungary, May 2011, pp. 1–5, 2011.

[17] L. Aladren, P. Garcia-Ducar, J. Mingo *et al.* (2011) Performance comparison of training sequences for power amplifier linearization systems. Digest 2011 IEEE International Symposium on Wireless Communication Systems (ISWCS), Aachen, Germany, November 2011, pp. 6–10, 2011.

[18] Saied-Bouajina, S., Hammi, O., Jaidane-Saidane, M. and Ghannouchi, F.M. (2010) Experimental approach for robust identification of radiofrequency power amplifier behavioural models using polynomial structures. *IET Microwaves, Antennas & Propagation*, **4** (11), 1418–1428.

[19] Hammi, O., Carichner, S., Vassilakis, B. and Ghannouchi, F.M. (2008) Synergetic crest factor reduction and baseband digital predistortion for adaptive 3G Doherty power amplifier linearizer design. *IEEE Transactions on Microwave Theory and Techniques*, **56** (11), 2602–2608.

[20] Hammi, O., Carichner, S., Vassilakis, B. and Ghannouchi, F.M. (2008) Power amplifiers' model assessment and memory effects intensity quantification using memoryless post-compensation technique. *IEEE Transactions on Microwave Theory and Techniques*, **56** (12), 3170–3179.

[21] Liu, T., Boumaiza, S. and Ghannouchi, F.M. (2005) Deembedding static nonlinearities and accurately identifying and modeling memory effects in wide-band RF transmitters. *IEEE Transactions on Microwave Theory and Techniques*, **53** (11), 3578–3587.

[22] T. Liu, Y. Ye, X. Zeng, and F.M. Ghannouchi, Accurate time-delay estimation and alignment for RF power amplifier/transmitter characterization. Proceedings 2008 IEEE International Conference on Circuits and Systems for Communications (ICCSC), Shanghai, China, May 2008, pp. 70–74, 2008.

[23] Hammi, O., Ghannouchi, F.M. and Vassilakis, B. (2008) On the sensitivity of RF transmitters' memory polynomial model identification to delay alignment resolution. *IEEE Microwave and Wireless Components Letters*, **18** (4), 263–265.

[24] O. Hammi, M. Younes, B. Vassilakis, and F.M. Ghannouchi, Digital predistorters sensitivity to delay alignment resolution. Digest 2009 IEEE Radio Wireless Symposium (RWS), San Diego, CA, January 2009, pp. 606–609, 2009.

[25] T. Liu, S. Boumaiza, M. Helaoui, *et al.* Behavior modeling procedure of wideband RF transmitters exhibiting memory effects. Digest IEEE MTT-S International Microwave Symposium (IMS'2005), Long Beach, CA, June 2005, pp. 1983–1986, 2005.

[26] T. Liu, Dynamic behavioral modeling and nonlinearity precompensation for broadband transmitters. PhD thesis. University of Montreal, 2005.

[27] H. Ben Nasr, S. Boumaiza, M. Helaoui, *et al.* On the critical issues of DSP/FPGA mixed digital predistorter implementation. Proceedings 2005 Asia Pacific Microwave Conference (APMC'2005), Suzhou, China, December 2005, pp. 1–4, 2005.

[28] A.K.C. Kwan Linearization of RF power amplifiers using digital predistortion technique implemented on a DSP/FPGA platform. Master of Science thesis. University of Calgary, 2009.

[29] Kwan, A., Helaoui, M., Boumaiza, S. *et al.* (2009) Wireless communications transmitter performance enhancement using advanced signal processing algorithms running in a hybrid DSP/FPGA platform. *The Journal of VLSI Signal Processing Systems for Signal, Image, and Video Technology*, **56** (2), 187–198.

[30] Moore, E.H. (1920) On the reciprocal of the general algebraic matrix. *Bulletin of the American Mathematical Society*, **26** (9), 394–395.

[31] Penrose, R. (1955) A generalized inverse for matrices. *Proceedings of the Cambridge Philosophical Society*, **51**, 406–413.

[32] Stoer, J. and Bulirsch, R. (2002) *Introduction to Numerical Analysis*, 3rd edn, Springer-Verlag, Berlin, New York.

[33] Haykin, S. (2002) *Adaptive Filter Theory*, 4th edn, Prentice Hall.

[34] Widrow, B. and Stearns, S.D. (1985) *Adaptive Signal Processing*, Prentice-Hall, Upper Saddle River, NJ.

[35] J. G. Proakis and D. G. Manolakis, *Digital Signal Processing: Principles, Algorithms and Applications*, Chapter 12, 3rd edn, Macmillan, 1996.

[36] Raich, R. and Zhou, G.T. (2004) Orthogonal polynomials for complex Gaussian processes. *IEEE Transactions on Signal Processing*, **52** (10), 2788–2797.

[37] Raich, R., Qian, H. and Zhou, G.T. (2004) Orthogonal polynomials for power amplifier modeling and predistorter design. *IEEE Transactions on Vehicular Technology*, **53** (5), 1468–1479.

[38] Helaoui, M., Boumaiza, S., Ghazel, A. and Ghannouchi, F.M. (2006) Power and efficiency enhancement of 3G multicarrier amplifiers using digital signal processing with experimental validation. *IEEE Transactions on Microwave Theory and Techniques*, **54** (4), 1396–1404.

9

Baseband Digital Predistortion

9.1 The Predistortion Concept

The ultimate aim of studying power amplifier (PA) distortions is to design appropriate predistorters that will compensate for these distortions and ensure linear amplification of the signal to be transmitted. In Chapter 3, the similarities between behavioral modeling and digital predistortion (DPD) were briefly introduced. Various mathematical formulations that can be used to implement behavioral models as well as digital predistorters were thoroughly discussed in Chapters 4–7. Chapter 8 exposed the common steps of the behavioral modeling and DPD processes with a focus on the identification techniques employed for the synthesis of the model or predistorter function. In this chapter, the specificities of DPD are addressed. Although the analysis is carried out in this chapter for the case of a PA, the concepts and results still hold in the case where the transmitter's analog front end is part of the device under test (DUT) to be linearized.

Conceptually, predistortion consists in implementing a nonlinear function upstream of the PA complementary to that of the amplifier to be linearized. Accordingly, the cascade made of the predistorter and the PA will operate as a linear amplification system as illustrated in the simplified block diagram of Figure 9.1. This figure also depicts sample amplitude modulation to amplitude modulation (AM/AM) and amplitude modulation to phase modulation (AM/PM) characteristics of the predistorter, the PA and the linearized power amplifier (LPA). The objective is to have a constant complex gain over the entire operating power range of the linearized amplifier. The power transfer characteristics of the predistorter, the PA and the LPA of Figure 9.1 are reported in Figure 9.2. This latter figure clearly illustrates that the predistorter is designed to generate a gain expansion that will compensate for the gain compression commonly observed in PAs. Since some class AB PAs as well as Doherty amplifiers tend to exhibit a gain expansion followed by a gain compression in their AM/AM characteristics, the predistorter has to compensate for these and thus must produce a gain compression followed by a gain expansion.

Behavioral Modeling and Predistortion of Wideband Wireless Transmitters, First Edition.
Fadhel M. Ghannouchi, Oualid Hammi and Mohamed Helaoui.
© 2015 John Wiley & Sons, Ltd. Published 2015 by John Wiley & Sons, Ltd.

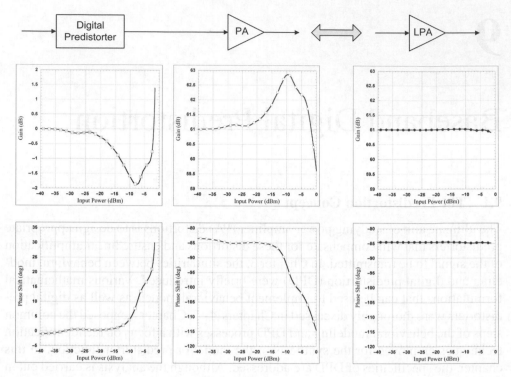

Figure 9.1 Simplified block diagram of predistortion system and corresponding gain characteristics of each block

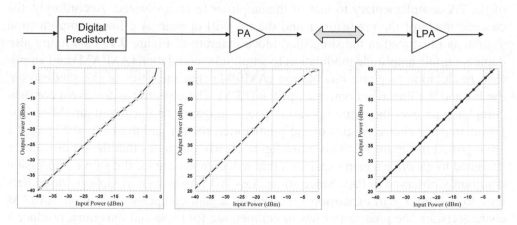

Figure 9.2 Power transfer characteristics involved in a predistortion system

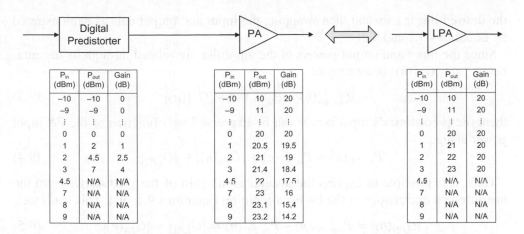

P_in (dBm)	P_out (dBm)	Gain (dB)
−10	−10	0
−9	−9	0
⋮	⋮	⋮
0	0	0
1	2	1
2	4.5	2.5
3	7	4
4.5	N/A	N/A
7	N/A	N/A
8	N/A	N/A
9	N/A	N/A

P_in (dBm)	P_out (dBm)	Gain (dB)
−10	10	20
−9	11	20
⋮	⋮	⋮
0	20	20
1	20.5	19.5
2	21	19
3	21.4	18.4
4.5	22	17.5
7	23	16
8	23.1	15.4
9	23.2	14.2

P_in (dBm)	P_out (dBm)	Gain (dB)
−10	10	20
−9	11	20
⋮	⋮	⋮
0	20	20
1	21	20
2	22	20
3	23	20
4.5	N/A	N/A
7	N/A	N/A
8	N/A	N/A
9	N/A	N/A

Figure 9.3 Power transfer characteristics in predistortion systems

A rudimentary numerical example that illustrates the predistortion concept is given in Figure 9.3 in which the input and output powers as well as the gains of a nonlinear PA and its predistorter are presented. This figure also includes the input and output power levels as well as the gain of the resulting linearized PA. In this case, the PA is assumed to have a small signal gain of 20 dB and a saturation output power in the range of 23 dBm. Even though this example considers a memoryless PA that does not cause AM/PM distortions, the same concept can be extended to include phase distortions and memory effects. As shown in Figure 9.3, a gain compression is observed in the PA characteristic for input power levels of 1 dBm and above. To compensate for this, the predistortion function introduces a complementary gain expansion. It is important to note here that the maximum input power of the predistorter is 3 dBm while the maximum input power of the PA is 7 dBm. Typically, when the input power of the DPD exceeds 3 dBm, clipping will occur in order to avoid overdriving the PA. The saturation input power of the DPD is determined by the saturation output power, or equivalently, the saturation input power of the PA and the gain of the linearized PA. This aspect will be further discussed in the DPD normalization gain section.

To ensure linear amplification, for a given output power of the PA at instant n ($P_{out\_PA}(n)$), the required power level at the input of the predistorter ($P_{in\_PD}(n)$) can be determined by:

$$P_{in\_PD}(n) = P_{out\_PA}(n) - |G_{LPA}| \qquad (9.1)$$

where G_{LPA} is the desired complex power gain of the linearized amplifier.

The output power of the predistorter ($P_{out\_PD}(n)$) is simply the input power of the PA ($P_{in\_PA}(n)$):

$$P_{out\_PD}(n) = P_{in\_PA}(n). \qquad (9.2)$$

Thus, the input-output power characteristic of the predistorter can be easily obtained from that of the PA by normalizing the output gain of the amplifier using the desired linear gain and then swapping the input and output data as demonstrated by Equations 9.1 and 9.2.

Since the input and output powers of the amplifier are related through its instantaneous gain $(G_{PA}(n))$ according to:

$$P_{out\_PA}(n) = P_{in\_PA}(n) + |G_{PA}(n)| \tag{9.3}$$

then, the predistorter's input power can be expressed as a function of the PA input power using:

$$P_{in\_PD}(n) = P_{in\_PA}(n) + |G_{PA}(n)| - |G_{LPA}|. \tag{9.4}$$

Thus, it is possible to express the instantaneous gain of the predistorter from the measured characteristics of the PA by combining Equations 9.1–9.4. This leads to:

$$|G_{PD}(n)| = P_{out\_PD}(n) - P_{in\_PD}(n) = |G_{LPA}| - |G_{PA}(n)|. \tag{9.5}$$

When AM/PM distortions are present, the phase distortions caused by the predistorter $\left(\left\lfloor G_{PD}(n) \right\rfloor\right)$ are:

$$\left\lfloor G_{PD}(n) \right\rfloor = \left\lfloor G_{LPA} \right\rfloor - \left\lfloor G_{PA}(n) \right\rfloor \tag{9.6}$$

$\left\lfloor G_{LPA} \right\rfloor$ and $\left\lfloor G_{PA}(n) \right\rfloor$ are the AM/PM distortions of the linearized PA and the PA, respectively.

In summary, once the AM/AM and AM/PM characteristics of the amplifier are measured, the AM/AM and AM/PM characteristics of the corresponding predistorter can be determined using Equations 9.4, 9.5, and 9.6. Having these desired predistorter characteristics, the models described in the previous chapters can be applied to accurately fit this dataset.

9.2   Adaptive Digital Predistortion

The effectiveness of DPD systems in canceling the distortions present at the output of PAs and transmitters extensively depends on the match between the predistorter's nonlinear characteristics and that of the DUT to be linearized. Since the nonlinearity exhibited by the DUT varies with time due to changes in the drive signal, aging, or drifts, it is essential to continuously update the predistortion function to maintain the linear operation of the system made of the predistorter and the DUT. Adaptive digital predistorters can be implemented either in closed loop or open loop configuration. This classification depends on the location of the predistortion function with respect to the adaptation loop.

9.2.1   Closed Loop Adaptive Digital Predistorters

In closed loop adaptive digital predistorters, the predistortion function is located inside the adaptation loop used to update the predistortion function coefficients.

The functional block diagram of closed loop DPD systems is depicted in Figure 9.4. The signal at the input of the digital predistorter $(x_{in\_DPD}(n))$ and that at the output of

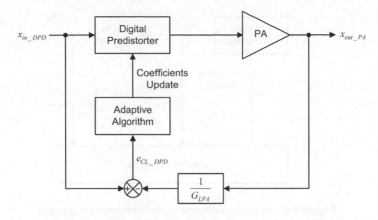

Figure 9.4   Closed loop adaptive digital predistortion system

the PA ($x_{out\_PA}(n)$) are used to compute the error signal of the closed loop DPD system ($e_{CL\_DPD}(n)$) defined by:

$$e_{CL\_DPD}(n) = x_{in\_DPD}(n) - \frac{x_{out\_PA}(n)}{G_{LPA}} \tag{9.7}$$

where G_{LPA} is the gain of the linearized PA.

The adaptive algorithm is then used to minimize this error signal and ensure that the amplifier's output signal is a scaled replica of the predistorter's input signal.

This concept is also known as "model the reference adaptive system" (MRAS) in control theory. Closed loop DPD systems employ the direct learning technique to identify the predistorter's coefficients. The direct learning refers to the method used to update the predistorter's coefficients by considering the input and output signals of the linearized PA made of the cascade of the predistorter and the amplifier [1, 2]. Closed loop predistorters usually exhibit slow convergence and high computational complexity since there is no direct relation between the error signal and the predistorter's coefficients. They are also prone to divergence if the PA is driven into saturation, which will cause the adaptive algorithm to repeatedly and unsuccessfully try to increase the output power of the predistorter to correct for the uncorrectable saturation induced distortions.

9.2.2   Open Loop Adaptive Digital Predistorters

DPD can be made adaptive without having to encompass the digital predistorter within the adaptation loop. Such systems are commonly referred to as open loop adaptive digital predistorters. A basic block diagram of open loop DPD systems is illustrated in Figure 9.5. In the open loop adaptive DPD system, the signals at the input and output of the PA are used to compute the DPD function update. This gives rise to two possible alternatives. In the first, the amplifier's model is identified and then the DPD function

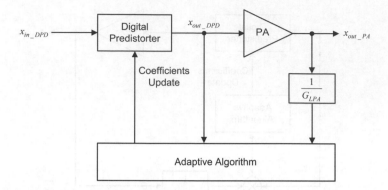

Figure 9.5 Open loop adaptive digital predistortion system

is built by inverting this model. Whereas in the second approach, the predistortion function is directly derived by calculating the post-inverse of the amplifier.

The two variants of open loop adaptive DPD systems are depicted in Figure 9.6. In Figure 9.6a, a direct learning scheme is used to identify the model of the PA. The inverse of this model, which represents the desired predistortion function, is then calculated. This technique is suitable for memoryless systems where one-to-one

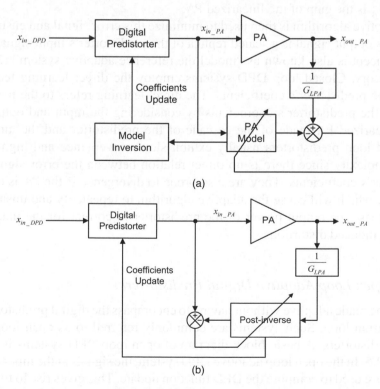

Figure 9.6 Open loop adaptive digital predistortion system implementations (a) direct learning architecture and (b) indirect learning architecture

mapping exists in the PA model. However, the presence of memory effects makes the amplifier model inversion very complicated and subject to substantial fitting errors.

To get around the need for calculating the inverse function of the PA model, the second variant of open loop adaptive DPD systems, illustrated in Figure 9.6b, identifies the post-inverse of the amplifier's nonlinearity by swapping its input and output signals. A copy of the post-inverse function is then used to predistort the input signal ($x_{in\_DPD}(n)$). The essence of this technique is based on the work of Schetzen on the theory of the pth order inverses of nonlinear systems where it was established that the pth order pre-inverse of a system is identical to its pth order post-inverse [3]. This is true under the assumption that up to the pth order nonlinear components are generated by the amplifier. This assumption is realistic and valid for PAs since high order nonlinear components have negligible contributions to the PA's behavior. This variant of open loop adaptive DPD, commonly used in adaptive digital predistorters, is referred to as the indirect learning technique [4, 5].

9.3 The Predistorter's Power Range in Indirect Learning Architectures

The major limitation in indirect learning architecture is that deriving the predistorter function by swapping the input and output data of the amplifier leads to a reduced power range of the predistorter when compared to that of the amplifier. To better illustrate this problem, let us consider a PA prototype to be linearized over its entire power range up to saturation. This analysis also holds if the amplifier is intended to be linearized only up to a certain output power lower than its saturation power. The power transfer characteristic of the considered PA prototype is reported in Figure 9.7. To simplify the calculations without restricting their validity, the PA's power transfer characteristic is reported after normalization by its small signal gain. The DPD's power transfer characteristic, as well as that of the linearized amplifier, is shown in the same figure. Here, the DPD was designed to have a 0 dB small signal gain and lead to a linearized amplifier gain equal to the PA's small signal gain ($|G_{LPA}| = |G_{SS\_PA}|$). In the remainder of this chapter, all gain and power values are expressed in dB and dBm units, respectively. The signal to be transmitted has a peak to average power ratio equal to $PAPR_{Sig}$.

It can be observed from the power transfer characteristics of the PA and its DPD that:

- The maximum power level at the predistorter's input is $P_{in\_max\_DPD}$. This power level can deduced from the PA's saturation point using:

$$P_{in\_max\_DPD} = P_{in\_sat\_PA} - GC_{sat\_PA} \qquad (9.8)$$

$P_{in\_sat\_PA}$ and GC_{sat_PA} are the PA's input power level and gain compression at saturation, respectively. These are shown in Figure 9.7.

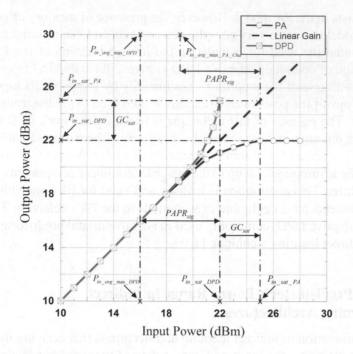

Figure 9.7   Power transfer characteristics of a PA and its predistorter

- The maximum average power level at the predistorter's input ($P_{in\_avg\_\max\_DPD}$). This represents the highest average power that can be applied at the DPD input without causing clipping. This power level depends on the signal's PAPR ($PAPR_{Sig}$) and is related to the predistorter's maximum power level ($P_{in\_\max\_DPD}$) by:

$$P_{in\_avg\_\max\_DPD} = P_{in\_\max\_DPD} - PAPR_{Sig}. \qquad (9.9)$$

Accordingly, the maximum average power that can be applied at the input of the PA during the linearization step is:

$$P_{in\_avg\_\max\_PA\_Lin} \simeq P_{in\_avg\_\max\_DPD} = P_{in\_sat\_PA} - GC_{sat\_PA} - PAPR_{Sig}. \qquad (9.10)$$

In this latter equation, it is assumed that the DPD does not introduce average power variations as mentioned in the case of the constant average power technique.

In contrast, to linearize the PA over its entire power range, its average input power during the characterization step should be:

$$P_{in\_avg\_\max\_PA\_Cha} = P_{in\_sat\_PA} - PAPR_{Sig}. \qquad (9.11)$$

Based on this, it appears that it is not possible to characterize the PA over its entire power range while operating it at the same average input power that will be applied during the linearization steps. To overcome this drawback of the indirect learning architecture, the constant average power technique, the constant peak power technique or the synergetic CFR (crest factor reduction)/DPD technique can be utilized.

9.3.1   Constant Peak Power Technique

One alternative that can be adopted to characterize the DUT over its entire power range is to adjust the power level of the input signal during the characterization step in order to drive the DUT over its entire power range. In this case, the average power of the DUT's input signal during the characterization step will be given by Equation 9.11. Conversely, during the linearization step, once the predistortion function is applied, the average power at the DUT's input will be equal to that given by Equation 9.10.

Combining Equations 9.8–9.11, it possible to estimate the average power variation at the input of the PA between the characterization and linearization steps. As a rule of thumb, this average power variation is expected to be in the range of the gain compression exhibited by the PA at peak power. In this case, since the PA is operated up to saturation,

$$P_{in\_avg\_PA\_Charac} - P_{in\_avg\_PA\_Lin} \simeq GC_{sat\_PA}. \qquad (9.12)$$

Such average power variation at the input of the PA will cause a change in its behavior. Consequently, a mismatch between the DPD nonlinearity and that of the PA will be noticed. Thus, iterative PA characterization and DPD synthesis procedures might be required to attain convergence, that is, the average power at the input of the PA remains quasi-unchanged between the characterization and the linearization steps.

This technique is referred to as the constant peak power technique since the PA is continuously driven up to its saturation power either during the characterization or the linearization steps.

9.3.2   Constant Average Power Technique

In this technique, the average power applied at the input of the predistorter during the characterization step (DPD OFF) is the same as that to be applied at the input of the predistorter during the linearization step (DPD ON). The concept is described here assuming that the predistorter's input and output signals have the same average power. A detailed study of the predistorter's gain effect of the system performance is reported in Section 9.4. This will establish that optimal performance is obtained when there is no average power variation through the predistorter.

The average power of the signal to be applied at the input of the predistorter during the linearization step was calculated in Equation 9.8. If during the characterization step the test signal is applied at the input of the DUT with an average power of $P_{in\_avg\_max\_DPD}$, the PA will not be characterized over its entire power range but only up to an input power level equal to $P_{in\_max\_DPD}$, which is several dBs lower than the PA's input saturation power (as previously derived in Equation 9.8). Consequently, the predistortion function derived from these measurements will not be defined over the required power range. To get around this shortcoming, the measured AM/AM and AM/PM characteristics of the PA can be extrapolated to approximate its behavior over the input power range spanning from $P_{in\_max\_DPD}$ to $P_{in\_sat\_PA}$. If needed, a second iteration of the PA characterization and DPD synthesis procedure can be performed.

In this iteration, and due to the gain expansion cause by the DPD, the PA will operate almost over its entire power range and thus the DPD function can be fully defined and the need for extrapolating the measured data will be alleviated.

Operating the amplifier at the same average power during the predistortion process ensures that its nonlinear characteristics remain quasi-unchanged and thus a perfect match between the DPD nonlinear characteristic and that of the PA will be obtained. The major drawback of this technique is related to the extrapolation of the PA data over the remainder of the power range. This extrapolation is straightforward in the case of look-up table models but becomes challenging in the presence of memory effects and the use of analytical models.

9.3.3   Synergetic CFR and DPD Technique

The constant peak power technique described above is not suitable for implementation in field deployed systems since it requires increasing the operating average power during the characterization step that is not realistic in such systems. On the other hand, the constant average power technique is viable for practical implementations but suffers from the extrapolation problem highlighted previously. The synergetic crest factor reduction and DPD technique was proposed, in [6], as a possible alternative to the constant average power technique that takes advantage of the co-existence of crest factor reduction and DPD modules in communication systems. The synergetic CFR/DPD technique is actually a constant average power technique in which the need for extrapolation is neatly dodged.

In this technique, the crest factor reduction is deliberately turned off during the PA characterization step. This allows for the identification of the predistortion function over the required power range. Then, during the linearization step, the crest factor reduction is turned on along with the DPD function. The thoughtful choice of using a high PAPR signal during the characterization step and a low PAPR signal during the linearization step eliminates the need for extrapolating the measured data as is required in the constant average power technique. The synergetic CFR/DPD technique can be perceived as a joint constant average power/constant peak power technique. In fact, the peak power at the input of the PA is almost maintained between the characterization and linearization steps due to the complementary actions of CFR and DPD on the signal's PAPR.

9.4   Small Signal Gain Normalization

The predistorter's AM/AM and AM/PM characteristics are derived from that of the DUT using Equations 9.4–9.6. These equations require the selection of the gain of the linearized amplifier G_{LPA}. This gain is applied to normalize the measured gain of the PA and define the small signal complex gain of the digital predistorter. Thus, this gain is referred to as normalization gain or linearized PA gain. The phase of the normalization gain is used to derive the AM/PM characteristic of the predistorter. The selection of the phase of the normalization gain is not critical for the performance of the predistorter as its variation only introduces a phase shift in the predistorted signal. Conversely, the choice of the magnitude of the normalization gain has a direct impact on predistorter performance. This is even more important when the AM/AM characteristic of the PA presents a gain expansion prior to the compression region or when the gain of the amplifier varies over a wide range as is the case of amplifiers biased in deep class AB.

Considering a typical power transfer characteristic of a PA, digital predistorters are commonly designed such that the linearized PA has the same gain as the PA. This implies that the DPD has a 0 dB small signal gain. This scenario is depicted in Figure 9.8, which shows the power transfer characteristics of a sample amplifier before and after linearization along with the AM/AM characteristic of the corresponding DPD.

For the study of the normalization gain importance in DPD, it is essential to focus on the following key power levels that are pointed out in Figure 9.8:

- $P_{in\_sat\_PA}$: the amplifier's input power level at saturation.
- $P_{out\_sat\_PA}$: the amplifier's output power level at saturation.
- $P_{in\_sat\_DPD}$: the predistorter's input power level at saturation.
- G_{SS_DPD}: the small signal gain of the predistorter.
- G_{sat_DPD}: the predistorter's gain at saturation.

Based on Equation 9.5, the magnitude of the predistorter's small signal gain is defined as:

$$|G_{SS\_DPD}| = |G_{LPA}| - |G_{SS\_PA}| \qquad (9.13)$$

where G_{SS_PA} is the amplifier's small signal gain.

The predistorter's input power and gain at saturation can be expressed as a function of the amplifier's output power and gain at saturation ($P_{out\_sat\_PA}$ and G_{sat_PA}, respectively) and the linearized amplifier gain (G_{LPA}) according to:

$$P_{in\_sat\_DPD} = P_{out\_sat\_PA} - |G_{LPA}| \qquad (9.14)$$

and

$$G_{sat\_DPD} = P_{in\_sat\_PA} - P_{in\_sat\_DPD} = |G_{LPA}| - |G_{sat\_PA}| \qquad (9.15)$$

The two latter equations describe how the AM/AM characteristics of the DPD evolve as the gain of the linearized amplifier is varied. Indeed, both $P_{in\_sat\_DPD}$ and G_{sat_DPD} can be controlled by adjusting the gain of the linearized amplifier (G_{LPA}). Thus, it is possible to design a plurality of predistorters having various power transfer characteristics by changing the normalization gain. The impact of the normalization gain on the DPD's AM/AM characteristics and the power transfer characteristic of the linearized amplifier is reported in Figure 9.9. This figure clearly shows that the maximum power

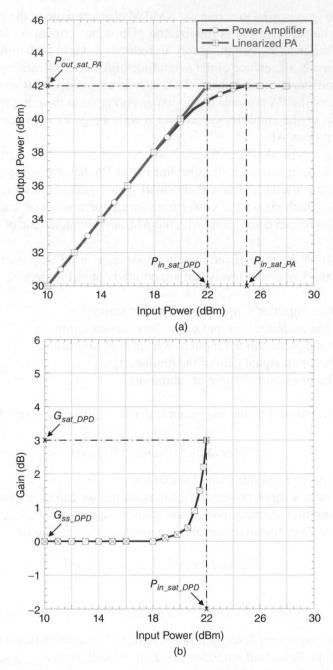

Figure 9.8 Characteristic curves of a power amplifier and its predistorter, (a) power transfer characteristic of the power amplifier before and after linearization and (b) AM/AM characteristic of the predistorter

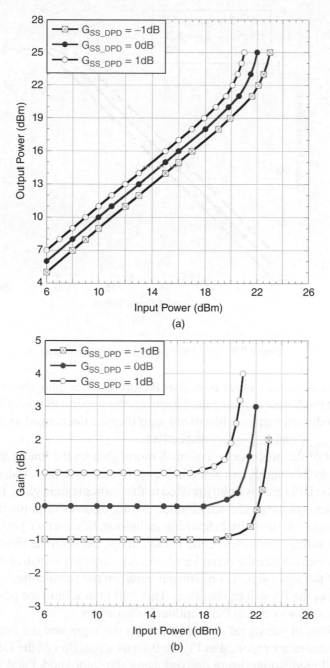

Figure 9.9   Impact of the normalization gain on the DPD and linearized DUT characteristics, (a) DPD's AM/AM characteristics, (b) DPD's power transfer characteristics

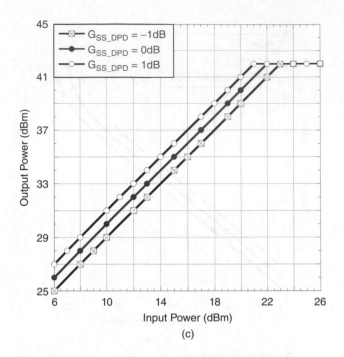

Figure 9.9   (c) LDUT power transfer characteristics

level at the predistorter's input as well as the predistorter gain characteristics vary as a function of the normalization gain. However, the maximum power levels at the output of the predistorter and the linearized amplifier are unchanged and are persistently equal to $P_{in\_sat\_PA}$ and $P_{out\_sat\_PA}$, respectively.

The impact of the predistorter's normalization gain on the linearized amplifier performances was thoroughly discussed in [7, 8]. In fact, as demonstrated in the analysis previously, the DPD gain is tightly related to the normalization gain. The presence of a power gain through the predistorter induces average power variation between its input and output signal. In the indirect learning technique, this average power variation will cause the operating conditions of the PA to change between the characterization and linearization steps. Since the behavior of PAs is commonly sensitive to average power variations of the input signal, a mismatch between the predistorter's nonlinear function and that of the PA will be observed. This will in turn limit the performance of the DPD and its ability to cancel the amplifier's distortions.

The variations of the signal's power between the input and the output of a sample digital predistorter are reported in Figure 9.10 as a function of the DPD's normalization gain. All predistorters were derived using the same model and parameters. The only difference is in the normalization gain used to derive the DPD characteristics. In this figure, each curve corresponds to a constant average power at the input of the predistorter. The variation of the signal's average power through the predistorter

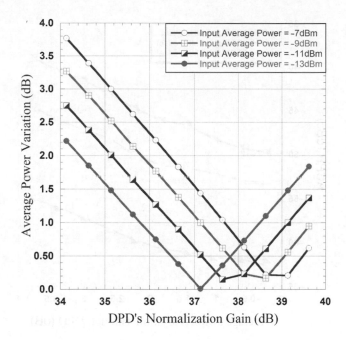

Figure 9.10 Effects of gain normalization on the average power variation through a digital predistorter

(ΔP_{avg_DPD}) is defined as:

$$\Delta P_{avg\_DPD} = |P_{out\_avg\_DPD} - P_{in\_avg\_DPD}| \qquad (9.16)$$

$P_{in\_avg\_DPD}$ and $P_{out\_avg\_DPD}$ correspond to the average power of the signals at the input and output of the predistorter, respectively.

The results presented in Figure 9.10 reveal that the average power variation through the predistorter can be controlled by appropriate selection of the normalization gain. For each operating average power, the optimal normalization gain is the one that leads to a minimal average power variation. Ideally, this average power variation should be 0 dB, however, in practice PA behavior is not sensitive to average power variations of up to 0.5 dB or even 1 dB in some cases.

The various digital predistorters derived in Figure 9.10 for the −7dBm average input power were applied to linearize the corresponding PA prototype. In these tests, the PA was driven by a two-carrier wideband code division multiple access (WCDMA) signal. The adjacent channel power ratios (ACPRs) at the output of a linearized PA prototype were measured for each digital predistorter. These measurement results are conveyed in Figure 9.11 as a function of the average power variation through the DPD. This figure confirms that the best DPD performance, corresponding to the lowest ACPR at the output of the linearized amplifier, is obtained when the average power

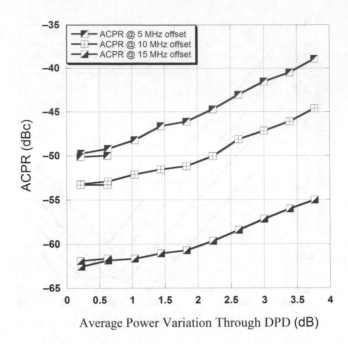

Figure 9.11   Effects of the average power variation through digital predistorter on the ACPR of a linearized power amplifier prototype [7]. ©2009 IEEE. Reprinted, with permission, from O. Hammi and F. M. Ghannouchi, "Power alignment of digital predistorters for power amplifiers linearity optimization," *IEEE Transactions on Broadcasting*, Mar. 2009

variation through the DPD is the lowest. Most importantly, this figure shows that significant deterioration in the DPD performance with up to 10 dB ACPR degradation can be observed as the average power variation through the DPD increases. For example, the spectra at the output of the same PA prototype were measured using two digital predistorters having identical models and parameters. The first DPD, referred to as conventional DPD, was derived by using the common practice according to which the normalization gain is equal to the PA's small signal gain ($G_{LPA} = G_{SS\_PA}$). Conversely, the second DPD, labeled 0 dB average power gain, was extracted by setting the normalization gain to its optimal value that minimizes the average power variation through the DPD. The spectra measured at the output of the PA using each of these predistorters as well as the one measured before applying the DPD are depicted in Figure 9.12. This corroborates the ACPR results and clearly displays the importance of the normalization gain optimization.

In order to counteract the DPD performance degradation following average power variation through the DPD, two approaches can be considered [7]. First, the DPD extraction procedure, including the PA characterization step, can be repeated iteratively until cancelation of the average power variation through the DPD occurs. Alternately, the DPD synthesis step can be repeated by adjusting the normalization

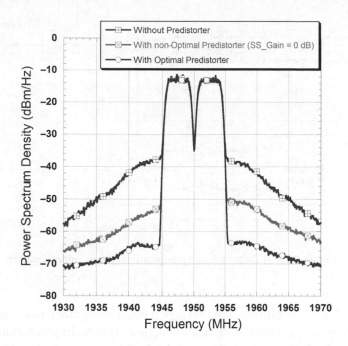

Figure 9.12   Effects of the DPD normalization gain on the output spectra of a linearized power amplifier prototype [7]. ©2009 IEEE. Reprinted, with permission, from O. Hammi and F. M. Ghannouchi, "Power alignment of digital predistorters for power amplifiers linearity optimization," *IEEE Transactions on Broadcasting*, Mar. 2009

gain in order to minimize the average power variation through the predistorter. A low complexity algorithm for the optimization of the DPD normalization gain was proposed in [8].

Varying the normalization gain of the DPD produces a change in the gain of the linearized DUT. Thus, it could be mistakenly expected that this would impact the power efficiency of the linearized PA. In fact, the normalization gain does not affect the PA's efficiency characteristics. Thus, the drain efficiency of the PA will remain unchanged. However, since the linearized amplifier's gain varies, the power added efficiency of the linearized PA will theoretically vary. Though, this variation is marginal [7]. For example, for a DUT having a drain efficiency of 50% and a gain of 30 dB; the power added efficiency variation is in the range of 0.2% if the normalization gain changes by 6 dB, which exceeds by far any reasonable variation of the normalization gain.

9.5   Digital Predistortion Implementations

9.5.1   Baseband Digital Predistortion

The predistortion concept for radio frequency (RF) PA linearization can be implemented either in an analog or a digital domain. With the continuous improvements in

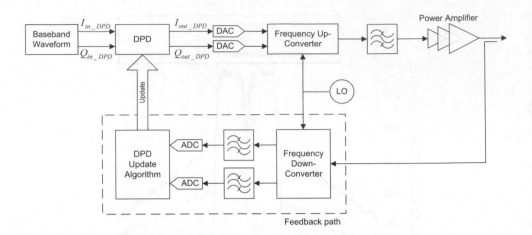

Figure 9.13   Baseband digital predistortion system implementation

terms of speed and capabilities of digital signal processing units, baseband DPD is the most commonly adopted predistortion architecture. Typical implementation of a baseband DPD system is illustrated in Figure 9.13. In this system, the baseband waveform is obtained at the output of a pulse shaping filter following data and source coding and mapping. The predistortion function is applied, in the digital domain, on this baseband waveform. The predistorted signal is then converted to an analog signal that will be up-converted and then fed to the power amplification stage. At the output of the PA, a coupler is used to feed back a fraction of the amplified signal. This feedback signal is down-converted, digitized, and then used to update the DPD function. This update can be done using either the direct learning or, most commonly, the indirect learning technique described in Section 9.2.

In the system implementation depicted in Figure 9.13, the input and output signals used to build the DPD function are the digital waveform at the predistorter output and the digitized version of the PA's output signal, respectively. This implies that any distortions or impairments present in the transmission path between the digital to analog converter and the output of the PA are part of the observation and can consequently be counteracted within the digital predistorter.

The popularity and widespread use of baseband DPD are mainly attributed to its flexibility in implementing any type of predistortion function with excellent accuracy. This makes it possible to meet the highly demanding linearity specifications of modern communication systems while using power efficient amplifiers. Indeed, when properly designed, baseband digital predistorters can compensate for the distortions of highly nonlinear amplifiers.

The feedback path maintains, ideally, a perfect match between the predistorter's and the PA's nonlinear functions. Moreover, the inclusion of memory effects using any of the models described in the previous chapters is straightforward since the baseband digital predistorter operates on the digital waveform. A full access to the complex samples is thus required, which might be an issue if the predistorter is needed for a system configuration where only the PA's RF (radio frequency) input and output signals are available.

The major limitation of baseband DPD is related to the bandwidth requirements of the signal generation and observation paths: Predistorting the signal to be transmitted results in bandwidth expansion. As a rule of thumb, one can expect the bandwidth of the predistorted signal to be five times wider than that of the signal before applying the predistortion function. Thus, baseband DPD sets severe bandwidth requirements on the digital to analog converters of the signal transmission path. Similar bandwidth requirements also apply to the analog to digital converters of the feedback path. However, this is not restricted to the case of baseband DPD and is also applied to the RF DPD.

In a research and development environment where the main task is to synthesize novel DPD functions or evaluate the linearizability of PA prototypes, the baseband DPD system of Figure 9.13 can be reproduced using a typical experimental setup that includes a vector signal generator and a vector signal analyzer as illustrated in Figure 9.14. The vector signal generator performs the functionalities of digital to analog converters as well as the RF front-end components of Figure 9.13 up to but excluding the power amplification system. These functionalities are digital to analog conversion, frequency up-conversion, and bandpass filtering. The vector signal analyzer is equivalent to the feedback path shown in Figure 9.13, and encompasses down-conversion and digitization steps. A computer is utilized to download the input signal waveforms into the vector signal generator and acquire the output signal waveforms from the vector signal analyzer. Software based algorithms are used to synthesize the DPD function and apply it to the input signal waveform to generate its predistorted version.

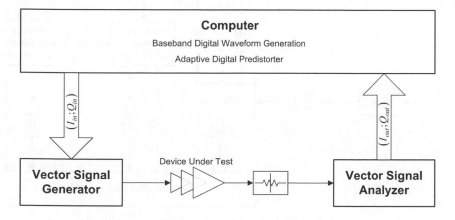

Figure 9.14   Measurement based digital predistortion test bed

9.5.2 RF Digital Predistortion

RF predistortion is a viable alternative to baseband DPD when the baseband digital waveform of the input signal is not accessible. RF predistortion is often implemented in an analog domain using diodes [9–12], or by controlling the complex gain of the driver stages [13]. In [14], a hybrid RF DPD system was introduced. In this system, the predistortion function is derived in the digital domain and applied on the RF analog signal through a vector modulator. This mixture between the analog and digital domain confers to this technique its hybrid character. As illustrated in Figure 9.15, the RF input and output signals of the PA are acquired through two identical down-conversion and digitization paths. This provides access to the baseband complex waveforms corresponding to the RF bandpass signals present at the input and output of the PA. These signals are used to identify the predistortion function using the open loop post-inverse based technique. A copy of the predistortion function is uploaded in the DPD module. A fraction of the RF input signal is applied at the input of an envelope detector, which provides the DPD with the instantaneous envelope of the RF signal. The corresponding predistortion coefficients are then applied through the complex multiplier to the RF input signal. In the main path, a delay line is used to align the input signal with the corresponding correction coefficients and compensate for the delay through the envelope detection path and the predistorter.

The main constraint in the RF DPD technique lies in the fact that only the envelope information of the signal to be transmitted is available and not its complex value as it is the case in baseband DPD. This restricts the functions that can be applied in RF DPD to either a look-up table structure [15–17] or an envelope memory polynomial based function [18].

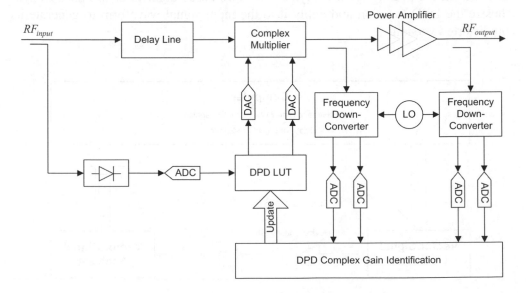

Figure 9.15 Hybrid RF digital predistortion system

9.6 The Bandwidth and Power Scalable Digital Predistortion Technique

As wireless communication systems evolve, the bandwidth of the signals to be handled by the PA and the RF front end keeps increasing. For example, in LTE-advanced (Long-Term Evolution-A) systems, the signal bandwidth can reach up to 100 MHz using carrier aggregation technique. This will emulate strong memory effects in the PA and calls for the use of DPD systems with a large number of coefficients. Coupled with the need for a fast predistortion function update due to the rapid changes in the bandwidth and power of signals to be transmitted, this situation calls for the development of appropriate DPD structures that are suitable for quick updates.

Based on the sensitivity of PA distortions to the characteristics of the drive signal, scalable digital predistorters have been proposed in [19]. These predistorters can be scaled with respect to the signal's average power and/or its bandwidth. In fact, as mentioned in Chapter 8, static distortions of PAs are mainly affected by the signal's average power, while the signal's bandwidth impacts the memory effects. This observation can be favorably applied in two-box based structures such as the twin-nonlinear two-box models described in Chapter 6 by making the memoryless sub-function of the model dependent on the signal's average power. The dynamic distortion sub-function of the model is then employed to track the variations of the signal bandwidth and is also used to compensate for residual distortions due to the mismatch between the memoryless nonlinear predistortion function and the DUT. Appropriate adaptation procedures can be applied to minimize the number of predistorter coefficients to be updated following changes in the drive signal characteristics.

Figure 9.16 presents a block diagram of the conventional reverse twin-nonlinear two-box based predistorter as well as its version proposed for implementing the power and bandwidth scalable digital predistorters as reported in [19]. The conventional reverse twin-nonlinear two-box model is considered as a single entity when it comes to the coefficients update, that is, the PA characterization data will be used to update both sub-functions of the model. Conversely, in the scalable version of this model, the update procedure is different. The memoryless nonlinear function that is commonly implemented using a LUT in the conventional reverse twin-nonlinear two-box model is replaced by a memoryless LUT bank in the scalable predistortion system. This memoryless LUT bank contains a plurality of memoryless LUTs that are indexed by the average power of the drive signal. The number of LUTs to be included in the LUT bank depends on the sensitivity of the PA's behavior to the variations of the signal's average power. However, this resolution is not critical for the performance of the predistorter as the residual nonlinearities will be removed by the dynamic distortion sub-functions of the predistorter.

The signals at the input of the memoryless nonlinear sub-function of the predistorter and at the output of the PA are used to derive the dynamic nonlinear function of the predistorter. This second box is frequently implemented using the memory polynomial model. The parameters of this sub-function, namely the nonlinearity order and

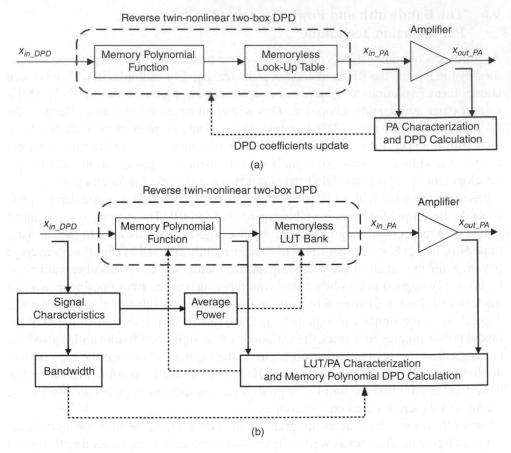

Figure 9.16 Digital predistortion system update. (a) Conventional approach and (b) bandwidth and power scalable approach

memory depth, are selected based on the bandwidth of the test signal. The placement of a dynamic nonlinear predistortion function is such that the signals used to derive it encompass the LUT predistorter and the nonlinear amplifier. Thus, it will compensate for all residual distortions due to possible mismatches between the static predistortion sub-function and the amplifier's nonlinear behavior. The bandwidth and power scalable DPD system was demonstrated to achieve the same performance as its conventional counterpart while requiring the update of up to 50% fewer coefficients [19].

9.7 Summary

In this chapter, the predistortion concept was introduced and various aspects related to the synthesis and implementation of digital predistortion functions were thoroughly discussed. The distinction between adaptive, open loop, and closed loop predistortion systems was illustrated and the advantages and drawbacks of each of these approaches highlighted. Then, the predistorter's power range issue observed in the widely adopted indirect learning approach was covered. It was shown that, to extend the power range over which the predistorter's function is defined, three approaches might be considered. The concept of each of these approaches was introduced and their pros and cons pointed out before a thorough analysis of the effects of the small signal gain normalization in the predistorter synthesis process was carried out.

Later in the chapter, the system level architectures of baseband, as well as RF digital predistorters, were presented. Scalability of digital predistortion systems, which ensures their fast and resource efficient adaptation in modern wireless application, was then discussed and an example of a system level concept that enables bandwidth and power scalability of the digital predistorter was exposed.

This chapter complements the previous chapters as it illustrates how the various models presented, as well as identification and characterization techniques, can be used to build the digital predistortion function and complete the process. It also sets the ground for the next chapter in which advanced topics related to multi-band and multiple-input multiple-output power amplification systems are presented.

References

[1] Gonzalez-Serrano, F.J., Murillo-Fuentes, J.J. and Artes-Rodriguez, A. (2001) GCMAC-based predistortion for digital modulations. *IEEE Transactions on Communications*, **49** (9), 1679–1689.

[2] Zhou, D. and DeBrunner, V.E. (2007) Novel adaptive nonlinear predistorters based on the direct learning algorithm. *IEEE Transactions on Signal Processing*, **55** (1), 120–133.

[3] Schetzen, M. (1976) Theory of pth-order inverses of nonlinear systems. *IEEE Transactions on Circuits and Systems*, **23** (5), 285–291.

[4] Eun, C. and Powers, E.J. (1997) A new Volterra predistorter based on the indirect learning architecture. *IEEE Transactions on Signal Processing*, **45** (1), 223–227.

[5] Ding, L., Zhou, G.T., Morgan, D.R. *et al.* (2004) A robust digital baseband predistorter constructed using memory polynomials. *IEEE Transactions on Communications*, **52** (1), 159–165.

[6] Hammi, O., Carichner, S., Vassilakis, B. and Ghannouchi, F.M. (2008) Synergetic crest factor reduction and baseband digital predistortion for adaptive 3G Doherty power amplifier linearizer design. *IEEE Transactions on Microwave Theory and Techniques*, **56** (11), 2602–2608.

[7] Hammi, O. and Ghannouchi, F.M. (2009) Power alignment of digital predistorters for power amplifiers linearity optimization. *IEEE Transactions on Broadcasting*, **55** (1), 109–114.

[8] Hammi, O., Boumaiza, S. and Ghannouchi, F.M. (2007) On the robustness of digital predistortion function synthesis and average power tracking for highly nonlinear power amplifiers. *IEEE Transactions on Microwave Theory and Techniques*, **55** (6), 1382–1389.

[9] P. Chan-Wang, F. Beauregard, G. Carangelo, and F. M. Ghannouchi, An independently controllable AM/AM and AM/PM predistortion linearizer for cdma2000 multicarrier applications. Proceedings 2001 IEEE Radio and Wireless Conference (RAWCON), Waltham, MA, August 2001, pp. 53–56, 2001.

[10] S. Rezaei, M. S. Hashmi, B. Dehlaghi, and F. M. Ghannouchi, A systematic methodology to design analog predistortion linearizer for dual inflection power amplifiers. Digest 2011 IEEE MTT-S International Microwave Symposium (IMS), Baltimore, MD, June 2011, pp. 1–4, 2011.

[11] Gupta, N., Tombak, A. and Mortazawi, A. (2004) A predistortion linearizer using a tunable resonator. *IEEE Microwave and Wireless Components Letters*, **14** (9), 431–433.

[12] Hu, X., Wang, G., Wang, Z.C. and Luo, J.R. (2011) Predistortion linearization of an X-Band TWTA for communications applications. *IEEE Transactions on Electron Devices*, **58** (6), 1768–1774.

[13] Son, K.Y., Koo, B. and Hong, S. (2012) A CMOS power amplifier with a built-in RF predistorter for handset applications. *IEEE Transactions on Microwave Theory and Techniques*, **60** (8), 2571–2580.

[14] E. G. Jeckeln, F. Beauregard, M. A. Sawan, and F. M. Ghannouchi, Adaptive baseband/RF predistorter for power amplifiers through instantaneous AM-AM and AM-PM characterization using digital receivers. Digest 2000 IEEE MTT-S International Microwave Symposium (IMS), Boston, MA, June 2000, pp. 489–492, 2000.

[15] Boumaiza, S., Jing, L., Jaidane-Saidane, M. and Ghannouchi, F.M. (2004) Adaptive digital/RF predistortion using a nonuniform LUT indexing function with built-in dependence on the amplifier nonlinearity. *IEEE Transactions on Microwave Theory and Techniques*, **52** (12), 2670–2677.

[16] Jeckeln, E.G., Ghannouchi, F.M. and Sawan, M.A. (2004) A new adaptive predistortion technique using software-defined radio and DSP technologies suitable for base station 3G power amplifiers. *IEEE Transactions on Microwave Theory and Techniques*, **52** (9), 2139–2147.

[17] Woo, W., Miller, M.D. and Kenney, J.S. (2005) A hybrid digital/RF envelope predistortion linearization system for power amplifiers. *IEEE Transactions on Microwave Theory and Techniques*, **53** (1), 229–237.

[18] Hammi, O., Ghannouchi, F.M. and Vassilakis, B. (2008) A compact envelope-memory polynomial for RF transmitters modeling with application to baseband and RF-digital predistortion. *IEEE Microwave and Wireless Components Letters*, **18** (5), 359–361.

[19] Hammi, O., Kwan, A. and Ghannouchi, F.M. (2013) Bandwidth and power scalable digital predistorter for compensating dynamic distortions in RF power amplifiers. *IEEE Transactions on Broadcasting*, **59** (3), 520–527.

10

Advanced Modeling and Digital Predistortion

In previous chapters, behavioral modeling and digital predistortion (DPD) have been discussed for conventional wireless transmitters consisting of a one-branch single-input single-output (SISO) transmitter having a signal in one frequency band. Such transmitter architectures do not cope with the increasing demand for high data rates. Therefore, the newest wireless communication standards propose the use of more advanced transmitters architectures that take advantage of the space diversity in a multi-input multi-output (MIMO) system in order to increase the transmission data rate; or that offer a better use of the wireless spectrum by transmitting concurrently in multiple bands for different standards (concurrent multi-standard transmission) or for the same standards (carrier-aggregated transmission). In both cases, multiple paths are considered in the power amplification systems, where each path has its own nonlinear characteristics, which makes the modeling of the nonlinear behavior more complex. In addition, the interaction between the paths makes the increase in complexity exponential and the modeling efforts much harder. The present chapter will address the efforts made in modeling and linearizing these advanced transmitter topologies. First, MIMO transmitters' models will be investigated. Then, the concurrent multi-band transmission will be modeled and solutions for its linearization will be presented.

10.1 Joint Quadrature Impairment and Nonlinear Distortion Compensation Using Multi-Input DPD

As described in Chapter 9, the DPD consists of placing a nonlinear module, the predistorter, in front of the radio frequency (RF) transmitter. This predistorter has a transfer function that is the inverse nonlinear response of the RF transmitter or power amplifier (PA). In order to extract the transmitter or PA nonlinear distortions, a feedback loop is

Behavioral Modeling and Predistortion of Wideband Wireless Transmitters, First Edition.
Fadhel M. Ghannouchi, Oualid Hammi and Mohamed Helaoui.
© 2015 John Wiley & Sons, Ltd. Published 2015 by John Wiley & Sons, Ltd.

needed to acquire the signal from the output of the PA, compare it to the input signal to the transmitter, and estimate the linear and nonlinear distortions. Therefore, the performance of the predistortion can be affected by linear distortions and imperfections in the transmitter and/or feedback paths. These linear distortions and imperfections are mainly due to the gain and phase imbalances and carrier leakages in quadrature modulator and demodulator. As a consequence, the inverse function of the PA cannot be accurately estimated and a significant residual error is obtained after linearization if these linear distortions are not taken into account.

These issues of PA linearization and quadrature imbalance compensation have been widely studied over the recent years. Until recently, most of the methods proposed thus far have addressed these two different problems separately [1–9]. Such an approach makes the proposed solution not practical to implement since access to both the quadrature modulator and the amplifier outputs are required. Only a few recently published papers have proposed models and techniques to address both issues jointly and simultaneously [10–14]. In the following, the quadrature imbalance problem and its effects on the quality of predistortion are investigated, first. Then, the multi-input models used for joint compensation of quadrature imbalance and DPD are summarized, and their performances analyzed and compared.

10.1.1 Modeling of Quadrature Modulator Imperfections

Quadrature imbalance is conventionally modeled using cross-coupled gains to account for the imbalance of the in-phase (I) and quadrature (Q) branches. Four filters h_{10}^I, h_{11}^I, h_{10}^Q, and h_{11}^Q in four different channels, composed of two straight I and Q channels and two cross-coupling channels, are used to model the quadrature imbalance [13]. Figure 10.1 shows the structure of the model for quadrature imbalance using four real filters. The baseband representation of the output signal from the modulator suffering from gain and phase imbalance can be expressed as follows:

$$y_{\text{mod}}(n) = y_{\text{mod}\_I}(n) + j y_{\text{mod}\_Q}(n) \tag{10.1}$$

where

$$y_{\text{mod}\_I}(n) = h_{10}^I x_I(n) + h_{11}^I x_Q(n) \tag{10.2}$$

$$y_{\text{mod}\_Q}(n) = h_{10}^Q x_Q(n) + h_{11}^Q x_I(n) \tag{10.3}$$

and $y_{\text{mod}\_I}(n)$ and $y_{\text{mod}\_Q}(n)$ are the in-phase and quadrature components of the signal at the output of the modulator, respectively; and $x_I(n)$ and $x_Q(n)$ are the in-phase and quadrature components of the baseband input signal, respectively.

The quadrature modulator baseband model can then be represented in its complex envelope format by:

$$y_{\text{mod}}(n) = h_I x_I(n) + h_Q x_Q(n) \tag{10.4}$$

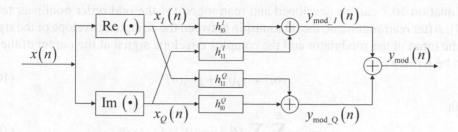

Figure 10.1 Modeling of the I/Q modulator's imbalance

where $h_I = h_{10}^I + jh_{11}^Q$ and $h_Q = h_{11}^I + jh_{10}^Q$ are the complex coefficients of the models that are applied to the in-phase and quadrature components of the input signal, respectively.

10.1.2 Dual-Input Polynomial Model for Memoryless Joint Modeling of Quadrature Imbalance and PA Distortions

If the bandwidth of the signal is small enough so that one can ignore the memory effects in the transmitter, the discrete-time equivalent baseband of the signal at the output of the modulator suffering from gain and phase imbalance can be expressed as:

$$y_{\text{mod}}(n) = h_{10}^I x_I(n) + h_{11}^I x_Q(n) + j[h_{10}^Q x_Q(n) + h_{11}^Q x_I(n)]. \qquad (10.5)$$

In the case of memoryless modeling, the common model used for PAs is the polynomial model [15, 16], where the output of the amplifier can be expressed as:

$$y(n) = \sum_{\substack{k=1 \\ k \text{ odd}}}^{K} H_k |y_{\text{mod}}(n)|^{k-1} y_{\text{mod}}(n). \qquad (10.6)$$

In Equation 10.6, $y_{\text{mod}}(n)$ is the baseband equivalent of the signal at the output of the quadrature modulator, which is the input to the PA; and $y(n)$ is the equivalent baseband of the signal at the output of the PA. K is the nonlinearity order of the amplifier's model and H_k represent the complex coefficients of the amplifier's model.

By replacing the expression of $y_{\text{mod}}(n)$ given in Equation 10.5, in the expression of the polynomial model of the PA given in Equation 10.6, the baseband equivalent of the signal at the output of the PA can be expressed as follows:

$$y(n) = \sum_{\substack{k=1 \\ k \text{ odd}}}^{K} H_k |h_{10}^I x_I(n) + h_{11}^I x_Q(n) + j[h_{10}^Q x_Q(n) + h_{11}^Q x_I(n)]|^{k-1}$$

$$\times \{h_{10}^I x_I(n) + h_{11}^I x_Q(n) + j[h_{10}^Q x_Q(n) + h_{11}^Q x_I(n)]\}. \qquad (10.7)$$

Equation 10.7 can be developed and rearranged for the odd order nonlinear terms [13]. After rearrangement, the relationship between the complex envelope of the signal at the input of the modulator and the complex envelope signal at the output of the PA can be written as:

$$y(n) = y_I(n) + jy_Q(n) \tag{10.8}$$

with

$$y_I(n) = \sum_{\substack{k=1 \\ k \, odd}}^{K} \sum_{r=0}^{k} H_{kr}^{I} [x_I(n)]^{k-r} [x_Q(n)]^{r} \tag{10.9}$$

and

$$y_Q(n) = \sum_{\substack{k=1 \\ k \, odd}}^{K} \sum_{r=0}^{k} H_{kr}^{Q} [x_Q(n)]^{k-r} [x_I(n)]^{r} \tag{10.10}$$

where H_{kr}^{I} and H_{kr}^{Q} are the real coefficients of the dual-input polynomial model related to the in-phase and quadrature components of the signal at the input of the RF front-end, respectively.

10.1.3 Dual-Input Memory Polynomial for Joint Modeling of Quadrature Imbalance and PA Distortions Including Memory Effects

Equations 10.9 and 10.10 include only the modeling of the PA's static nonlinearity and the static quadrature imbalance of the modulator. No memory effects are considered in this model. To extend this model and include the memory effects of both the PA and the quadrature modulator, the baseband equivalent model of the cascaded RF front-end including the PA and the quadrature modulator will then be given by:

$$y(n) = \sum_{m=0}^{M} \sum_{\substack{k=1 \\ k \, odd}}^{K} \sum_{r=0}^{k} H_{kr}(m) [x_I(n-m)]^{k-r} [x_Q(n-m)]^{r} \tag{10.11}$$

where $H_{kr}(m) = H_{kr}^{I}(m) + jH_{kr}^{Q}(m)$ are the complex coefficients of the dual-input memory polynomial model, M is the memory depth of this model and K is its nonlinearity order.

The polynomial model described by Equation 10.6 contains only nonlinearity terms of odd orders. It has been shown, however, that including even-order terms in this model reduces the modeling error and improves its accuracy for both forward and reverse modeling of PAs [17]. The dual-input memory polynomial model of Equation 10.11 can be modified to include even-order nonlinearity terms and to include a dc term H_{00}, which models the carrier leakage in the quadrature modulator. The final expression of the dual-input memory polynomial model is then:

$$y(n) = \sum_{m=0}^{M} \sum_{k=0}^{K} \sum_{r=0}^{k} H_{kr}(m) [x_I(n-m)]^{k-r} [x_Q(n-m)]^{r}. \tag{10.12}$$

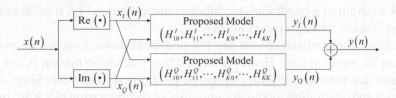

Figure 10.2 Dual-input nonlinear model for the joint effects of quadrature impairments and PA nonlinearities

This formulation is able to model concurrently the nonlinearity and memory effects introduced by the PA, as well as the dynamic gain and phase imbalance and the carrier leakage in the quadrature modulator. Figure 10.2 shows the block diagram of the dual-input memory polynomial model.

10.1.4 Dual-Branch Parallel Hammerstein Model for Joint Modeling of Quadrature Imbalance and PA Distortions with Memory

The quadrature impairments in terms of gain and phase imbalance, which were modeled in Section 10.1.1 using a set of four filters applied to the in-phase and quadrature components, can also be modeled by two complex filters, G_1 and G_2, where each of them takes a complex signal as input. The input to G_1 is the original or non-conjugate complex envelope signal to be fed to the quadrature modulator and the input to G_2 is the conjugate of this complex envelope signal [10, 11, 18, 19]. It can easily be shown that the inverse model (predistorter) of this system has a structure similar to that of the forward model.

The PA nonlinearity can be corrected with any of the predistortion functions presented in the previous chapters. In [11], a Hammerstein model is used. The joint compensation of the quadrature modulator impairments and PA nonlinearity can then be obtained by cascading the Hammerstein predistorter and the quadrature impairment compensation block, where the predistorter is placed first. Figure 10.3 shows

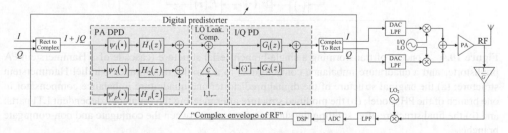

Figure 10.3 Block diagram of a linear and nonlinear distortion compensation composed of a cascade of a Hammerstein PA predistorter and a quadrature imbalance compensator [11]

the block diagram of a quadrature modulator impairment and PA nonlinear distortion compensation model as proposed in [11].

If the quadrature imbalance and the PA nonlinear distortions are known, such a model can be implemented. However, in practice, such information is not available beforehand and therefore a joint estimation is required. Since the filters of the PA predistorter and quadrature modulator impairment compensation block are in cascade, their joint estimation is not straightforward. In order to be able to achieve a joint estimation of the different coefficients of the model, the structure of Figure 10.3 can be modified from a cascade to a parallel structure, enabling one-step joint estimation of all the parameters using linear LS techniques, without any extra RF hardware.

This transformation can be achieved in different steps as described in [11]. First, for each branch of the parallel Hammerstein (PH) model, the quadrature compensation is added to the frequency response of that branch, as shown in Figure 10.4a. As shown in Figure 10.4b, the two frequency responses are merged together; $H_p(z)$ and $G_1(z)$ are merged in the non-conjugate path while $H_p^*(z)$ and $G_2(z)$ are merged together in

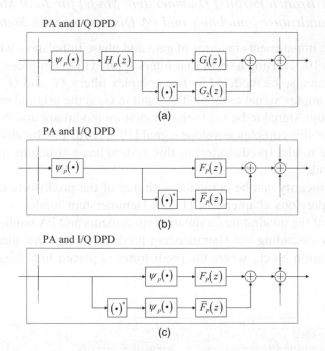

(a)

(b)

(c)

Figure 10.4 Concept of transforming a single-branch serial structure (cascade of a Hammerstein PA predistorter and a quadrature imbalance compensator) to a parallel dual-branch parallel Hammerstein structure: (a) the original structure of the digital predistorter and quadrature imbalance compensator in one branch of the PH model; (b) the modified structure after merging the frequency dependent LTI parts; and (c) the final structure after splitting the static nonlinearity between the conjugate and non-conjugate branches

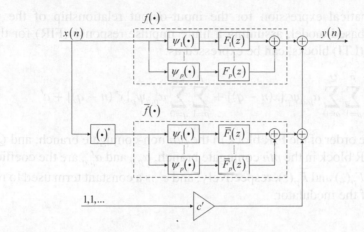

Figure 10.5   Detailed block diagram of a parallel dual-branch parallel Hammerstein structure

the conjugate path to obtain the following transfer functions:

$$F_p(z) = H_p(z)G_1(z) \tag{10.13}$$

$$\overline{F}_p(z) = H_p^*(z)G_2(z) \tag{10.14}$$

where $p \in I_P$, $I_P = \{1, 2, 3, \dots, P\}$ is the set of used polynomial orders, and P is the number of branches considered in the Hammerstein model.

Then, as depicted in Figure 10.4c, the final structure of the cascade obtained by splitting the static nonlinearity part of the predistorter function represented by a polynomial, $\psi_p[x(n)] = \sum_{k\in I_p} a_{kp}|x(n)|^{k-1}x(n)$, between the non-conjugate and a conjugate branches, resulting in a parallel connection of two PH predistorters with summed outputs. The time domain analysis of this model is explained in the following equations:

$$f_P[x(n)] = \sum_{p=1}^{P} f_{p,n} \otimes \psi_p[x(n)] \tag{10.15}$$

$$\overline{f}_P[x^*(n)] = \sum_{p=1}^{P} \overline{f}_{p,n} \otimes \psi_p[x^*(n)] \tag{10.16}$$

where $(.)^*$ is the complex conjugate operator; $\otimes$ is the convolution sum operator; and $f_{p,n}$ and $\overline{f}_{p,n}$ are the impulse responses of the transfer functions $F_P(z)$ and $\overline{F}_P(z)$, respectively.

Finally, the local oscillator (LO) leakage is compensated for by a constant added to the signal at the output of the joint linearization architecture. This constant is named c'; and the entire block diagram of the joint compensation becomes then as depicted in Figure 10.5.

The mathematical expression for the input-output relationship of the parallel-Hammerstein-based model, assuming a finite impulse response (FIR) for the linear time invariant (LTI) blocks, can be expressed as:

$$y(n) = \sum_{p=1}^{P} \sum_{q=0}^{Q_p} \alpha_{p,q} \psi_p[x(n-q)] + \sum_{p=1}^{P} \sum_{q=0}^{Q'_p} \alpha'_{p,q} \psi_p[x^*(n-q)] + c' \qquad (10.17)$$

where Q_p is the order of the FIR block in the *pth* non-conjugate branch, and $Q_p{}'$ is the order of the FIR block in the *pth* conjugate branch, $\alpha_{p,q}$ and $\alpha'_{p,q}$ are the coefficients of the FIR filters $F_p(z)$ and $\overline{F}_p(z)$, respectively; and c' is a constant term used to represent the dc offset of the modulator.

10.1.5 Dual-Conjugate-Input Memory Polynomial for Joint Modeling of Quadrature Imbalance and PA Distortions Including Memory Effects

In Equation 10.17, the function ψ_p represents the static nonlinearity and can be implemented either by a polynomial function, a look-up-table, or any memoryless model. The model described by Equation 10.17 can be generalized by using a memory polynomial model instead of the memoryless nonlinear function. Part of the system's memory is then compensated by this memory polynomial function while residual memory and quadrature imbalance distortion are compensated by the two filters in the non-conjugate and conjugate paths. The signal at the output of proposed model can then be written as follows:

$$y(n) = \sum_{p=1}^{P} \sum_{q=0}^{Q_p} \sum_{k=1}^{K} \alpha_{p,q,k} x(n-q)|x(n-q)|^{k-1}$$

$$+ \sum_{p=1}^{P} \sum_{q=0}^{Q'_p} \sum_{k=1}^{K} \alpha'_{p,q,k} x^*(n-q)|x^*(n-q)|^{k-1} + c' \qquad (10.18)$$

where $x(n)$, $x^*(n)$, and $y(n)$ are the input, conjugate of the input and output complex envelop signals, respectively; and c' is a constant term used to represent the dc offset of the modulator[12]. Figure 10.6 describes the concept of the dual-conjugate-input memory polynomial model of Equation 10.18.

10.2 Modeling and Linearization of Nonlinear MIMO Systems

10.2.1 Impairments in MIMO Systems

Crosstalk is the coupling effect between two or more signals sources. Such coupling results in interference between the different signals. In the case of MIMO

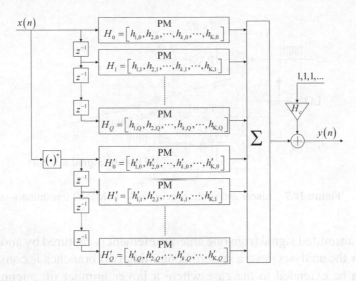

Figure 10.6 Detailed block diagram of the dual-conjugate-input memory polynomial structure

transmitters, crosstalk is the result of coupling between signals in separate paths. In MIMO configurations where the various paths use the same operating carrier frequency, crosstalk results in problematic interference between the paths. This crosstalk would be more significant in integrated circuit (IC) designs, where the physical distance between the signal paths is small and their perfect isolation is not possible.

One can categorize crosstalk in MIMO transmitters in two different categories: linear or antenna crosstalk and nonlinear crosstalk. The crosstalk is said to be linear if it can be modeled using linear functions of the interfering signals and the desired signal. In fact, if the signal affected by cross-coupling from other signal paths does not pass through a nonlinear system, the crosstalk can be modeled by a linear function and is considered as linear crosstalk. However, if the cross-coupled signal passes through a nonlinear system, the crosstalk includes nonlinear terms of the cross-coupling and can only be modeled using nonlinear equations. In this case, it is said nonlinear crosstalk. Since the PA is the main source of nonlinearity in wireless transmitters, crosstalk that occurs in the transmitter circuit before the PA is the main source of nonlinear crosstalk, while any crosstalk taking place after the PA is considered as linear crosstalk [20, 21]. Figure 10.7 shows a block diagram of a typical MIMO transmitter and illustrates the previously mentioned two types of crosstalk.

10.2.1.1 Linear or Antenna Crosstalk

Linear crosstalk in MIMO transmitters can be defined as the leakage between the different branches after the PAs. Most of this type of leakage can occur at the antenna as

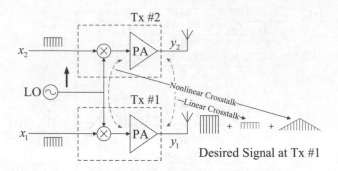

Figure 10.7   Linear and nonlinear crosstalks in MIMO transmitters

part of the transmitted signal from one antenna element is captured by another antenna element. For the analysis next, a system containing two branches is considered. This analysis can be extended to the case where a larger number of antennas are used. Linear crosstalk is usually modeled as follows:

$$\mathbf{v} = \mathbf{A}\mathbf{y} \tag{10.19}$$

where $\mathbf{y} = \begin{bmatrix} y_1 & y_2 \end{bmatrix}^T$ is a vector of the baseband equivalents, y_1 and y_2, of the two signals at the input of the transmitting antennas; $\mathbf{v} = \begin{bmatrix} v_1 & v_2 \end{bmatrix}^T$ is a vector of the baseband equivalents, v_1 and v_2, of the two signals at the output of the transmitting antennas; and $\mathbf{A} = \begin{bmatrix} 1 & \alpha \\ \beta & 1 \end{bmatrix}$ is a matrix modeling the antennas' linear crosstalk. In the case of a symmetric crosstalk, α and β are equal. However, when no crosstalk is present, $\alpha = \beta = 0$.

Equation 10.19 can be written in a linear system format as follows:

$$\begin{cases} v_1 = y_1 + \alpha y_2 \\ v_2 = \beta y_1 + y_2 \end{cases} \tag{10.20}$$

It can be observed from Equation 10.20 that the signals v_1 and v_2 obtained at the output of the first and second antenna, respectively, after crosstalk are linear functions of the input signals to the transmitting antennas, y_1 and y_2. The compensation for the effect of the antenna linear crosstalk is performed generally at the receiver side concurrently with the compensation of the linear crosstalk that is generated by the channel [22–25].

The composite linear crosstalk generated at the transmitter and receiver antennas and by the channel can be shown to have similar model. Figure 10.8 shows a block diagram illustrating a 2 × 2 MIMO system with three different types of linear crosstalk:

- The transmitting antennas crosstalk, modeled by the matrix $\mathbf{A}$,
- The channel crosstalk, modeled by the matrix $\mathbf{H} = \begin{bmatrix} h_{11} & h_{12} \\ h_{21} & h_{22} \end{bmatrix}$, and

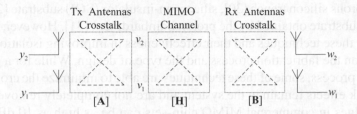

Tx Antennas   MIMO   Rx Antennas
Crosstalk   Channel   Crosstalk

Figure 10.8   Linear crosstalk model in MIMO configuration

- The receiving antennas crosstalk, modeled by the matrix **B**.

The overall linear crosstalk can then be modeled as follows:

$$\mathbf{w} = (\mathbf{BHA})\mathbf{y} + \mathbf{Bn} \tag{10.21}$$

where $\mathbf{w} = \begin{bmatrix} w_1 & w_2 \end{bmatrix}^T$ is a vector of the baseband equivalents, w_1 and w_2, of the two signals at output of the receiving antennas; and **n** is a vector that represents the additive white Gaussian noise (AWGN) in the different paths. The uncorrelated received signal, $\mathbf{w_{un}}$, can be obtained by inverting the matrix **BHA** as follows:

$$\mathbf{w_{un}} = (\mathbf{BHA})^{-1}\mathbf{w}$$

$$\mathbf{w_{un}} = \mathbf{y} + (\mathbf{HA})^{-1}\mathbf{n} \tag{10.22}$$

The uncorrelated received signal represents an estimation of the original signal **y** and an additional noise component. From Equation 10.22, it can be observed that:

- This method assumes that the matrix **BHA** is invertible; otherwise, the compensation for the linear crosstalk will not be possible.
- The crosstalks in the transmitter's antennas and in the MIMO channel may degrade the performance of the MIMO system since they increase the effect of noise as illustrated in the last term of Equation 10.22 [26].
- If the matrix **B** is invertible, the noise is not affected by the receiver's antennas crosstalk and so is the performance of the MIMO system.

10.2.1.2   Nonlinear Crosstalk

As defined at the beginning of this section, the nonlinear crosstalk in MIMO systems is generated by any cross-coupling between the MIMO paths that occurs before the PA. Such nonlinear crosstalk can be caused by several factors, the most important of them are the leakage of the RF signals through the common LO path due to non-perfect isolation of mixers [27], and the RF signals cross-coupling due to interferences in the chipset. Several techniques have been proposed in the literature to reduce this coupling such as buffering the LO paths [27], grounded guard ring [28], deep trench

[28, 29], porous silicon trench [30], silicon-on-insulator (SOI) substrate [28], and high resistivity substrate obtained by the proton bombardment [31]. However, the possibility to apply these techniques and their effectiveness in improving isolation are mainly dependent on the fabrication process and the type of design. While for a given design and a given process, some of these techniques are able to minimize the crosstalk, residual crosstalk effects remain in the system and are not completely removed. Crosstalk residual values in commercial MIMO chip-sets can be as high as 10 dB. Therefore, modeling this phenomenon and analyzing its effects on signal quality and on the linearization process are of significant importance.

If the transmitters do not include predistortion, as shown in Figure 10.7, their outputs, with the effect of crosstalk, can be modeled as:

$$y_1 = f_1(x_1 + \alpha x_2)$$
$$y_2 = f_2(\beta x_1 + x_2) \qquad\qquad (10.23)$$

where x_1 and x_2 are the MIMO transmitter inputs, y_1 and y_2 are the MIMO transmitter outputs, and $f_1(\cdot)$ and $f_2(\cdot)$ are functions representing the nonlinear responses of each of the transmitter's two branches. Since these functions are nonlinear, this crosstalk is said to be nonlinear and thus matrix inversion technique cannot be used to compensate for it. Moreover, it was shown, in [21], that this nonlinear crosstalk impacts the quality of the predistortion algorithm since the complex envelope of each RF signal at the input of the PAs is changed by the cross-coupling [21].

Figure 10.9 shows a block diagram of a MIMO transmitter that has nonlinear crosstalk and that uses DPD to compensate for the nonlinearity of the PAs. In the first path, the coefficients of the DPD function, $g_1(\cdot)$, are extracted using the digital baseband signal at the input of the first transmitting branch, z_1, taken at the output of

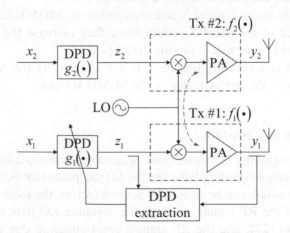

Figure 10.9 Conventional digital predistortion in the presence of nonlinear crosstalk in a MIMO transmitter

the DPD function and the complex envelope equivalent signal, y_1, at the output the first branch taken at the output of its PA. Similarly, in the second path, the coefficients DPD function, $g_2(\cdot)$, are estimated using the complex baseband waveforms at the input and output of this branch (z_2 and y_2, respectively). Since the nonlinear crosstalk occurs before the nonlinear response of each path, prior to the amplification, the output of the PA in each path is a nonlinear function of both inputs and can be expressed by replacing in Equation 10.23 the input signals (x_1 and x_2) by their predistorted versions (z_1 and z_2, respectively) to take into account the presence of the DPD functions (g_1 and g_2, respectively). Accordingly,

$$y_1 = f_1(z_1 + \alpha z_2)$$

$$y_2 = f_2(\beta z_1 + z_2) \, . \tag{10.24}$$

If the predistortion function $g_1(\cdot)$ was exactly the inverse of $f_1(\cdot)$, then it should be dependent on the two signals, x_1 and x_2 since the output of $f_1(\cdot)$ is function of z_1 and z_2 simultaneously. However, for conventional predistortion models as were presented in the previous chapters, the input signal of $g_1(\cdot)$ is only a function of x_1. As a result, the signal at the output of the first branch's PA can be written as:

$$y_1 = f_1[g_1(x_1) + \alpha z_2] \neq G_1 x_1 \tag{10.25}$$

and, similarly, the signal at output of the PA of the transmitter's second branch can be written as:

$$y_2 = f_2[g_2(x_2) + \alpha z_1] \neq G_2 x_2 \tag{10.26}$$

where G_1 and G_2 are the small signal gain of the transmitter's first and second branches.

Thus, using the previously mentioned linearization approach, the output of each branch cannot be only function of its corresponding input, and therefore no accurate linearization can be achieved. One can conclude that the nonlinear crosstalk deteriorates the performance of conventional DPD and a better model should be used for the modeling and linearization of MIMO systems in which nonlinear crosstalk is present. In the remaining part of this section, most important MIMO transmitters' behavioral models that account for nonlinear crosstalk will be discussed.

10.2.2   Crossover Polynomial Model for MIMO Transmitters

In the previous chapters, it was shown that the nonlinearity of PAs can be modeled using a polynomial model in the memoryless case and a memory polynomial to compensate for nonlinear distortions and memory effects. For a given PA, the pass-band memoryless polynomial model can be written as:

$$\tilde{y}(t) = \sum_{k=1}^{K} h_k \tilde{x}^k(t) \tag{10.27}$$

where $\tilde{x}(t)$ and $\tilde{y}(t)$ are the pass-band signals at the input and output of the PA, respectively; and K is the polynomial order. The baseband equivalent of this model relating the complex envelope of the input signal, $x(n)$, to the complex envelope of the output signal, $y(n)$, is given in Equation 4.50 and is provided here for convenience:

$$y(n) = \sum_{k=1}^{K} a_k |x(n)|^{k-1} x(n). \qquad (10.28)$$

In the case of MIMO system with nonlinear crosstalk, in order to take into account the dependence of each PA nonlinear behavior on the input signals of different paths, the memoryless polynomial model of Equation 10.28 can be extended so that the output of each of the two branches is the summation of different polynomial functions of each input. In the case of a dual-input system, the crossover model is then given as follows [21]:

$$y_1(n) = \sum_{k=1}^{K_{11}} a_{11,k} |x_1(n)|^{k-1} x_1(n) + \sum_{k=1}^{K_{12}} a_{12,k} |x_2(n)|^{k-1} x_2(n)$$

$$y_2(n) = \sum_{k=1}^{K_{21}} a_{21,k} |x_1(n)|^{k-1} x_1(n) + \sum_{k=1}^{K_{22}} a_{22,k} |x_2(n)|^{k-1} x_2(n) \qquad (10.29)$$

where K_{11}, K_{12}, K_{21}, and K_{22} represent the nonlinearity orders of the crossover model; and $a_{11,k}$, $a_{12,k}$, $a_{21,k}$, and $a_{22,k}$ are the crossover model's coefficients.

While each output in the crossover model is a function of both input signals $x_1(n)$ and $x_2(n)$, this model makes the approximation that there is no effect of one input signal on the nonlinear function applied to the other signal, which is not true in practice. The crossover model lacks the ability to predict any nonlinear cross-term, which results in sub-optimal modeling and linearization performances.

10.2.3 Dual-Input Nonlinear Polynomial Model for MIMO Transmitters

In order to take into account the effect of one branch's input signal on the nonlinear function applied to the other branch's input signal, the polynomial model should include nonlinear cross-terms of the two signals. A tensor product is used in the dual-input nonlinear polynomial model to create these cross-terms. The equation for this model in the pass-band format is given by:

$$\tilde{\mathbf{y}}(t) = \mathbf{h}_1 \tilde{\mathbf{x}}(t) + \mathbf{h}_2 [\tilde{\mathbf{x}}(t) \otimes \tilde{\mathbf{x}}(t)] + \cdots + \mathbf{h}_K [\tilde{\mathbf{x}}(t) \otimes \cdots \otimes \tilde{\mathbf{x}}(t)] \qquad (10.30)$$

where $\tilde{\mathbf{x}}(t) = \begin{bmatrix} \tilde{x}_1(t) \\ \tilde{x}_2(t) \end{bmatrix}$, $\tilde{x}_1(t)$, and $\tilde{x}_2(t)$ are the pass-band input signals of the first and second branches, respectively; $\mathbf{h_k} = [h_{k,1}, \ h_{k,2}, \ \ldots \ \ldots \ h_{k,2^k}]$ where $h_{k,j}$ represent the model coefficients, and $\otimes$ is used to denote the Kronecker product defined as:

$$\begin{bmatrix} a_1 \\ a_2 \end{bmatrix} \otimes \begin{bmatrix} b_1 \\ b_2 \end{bmatrix} = \begin{bmatrix} a_1 b_1 \\ a_1 b_2 \\ a_2 b_1 \\ a_2 b_2 \end{bmatrix}. \tag{10.31}$$

In Equation 10.30, $\tilde{y}(t)$ refers to the pass-band output of one of the two branches. Because of symmetry, one can only analyze the output of one branch, without loss of generality. For easiness of notation of the output signal, the index of the branch is dropped in the remaining analyses of the MIMO system.

This model includes all the cross-terms between $\tilde{x}_1$ and $\tilde{x}_2$. Therefore, the dual-input nonlinear polynomial is able to model the effects of nonlinear crosstalk more accurately along with the PA's nonlinearity. Although the accuracy of the model is improved, the complexity of the problem is dramatically increased as the number of model coefficients increases exponentially with the polynomial order. Moreover, it can be easily shown that because of the Kronecker product, many cross-terms are repeated, which results in unnecessary complexity increase.

10.2.4 MIMO Transmitters Nonlinear Multi-Variable Polynomial Model

In order to reduce the complexity of the model described in the previous section, a tradeoff between modeling accuracy and complexity should be considered. The nonlinear multi-variable polynomial model was proposed to achieve such trade-off [32]. This multi-variable polynomial model maintains all the possible cross-term products between the different input signals. It reduces the complexity of the dual-input polynomial model by removing the duplicated terms generated by the Kronecker product. For example, when $k = 2$, the cross-terms generated by the second order nonlinearity are $\tilde{x}_1 \tilde{x}_2$ and $\tilde{x}_2 \tilde{x}_1$. Even though these terms have different coefficients in the dual-input nonlinear polynomial model, they are in fact similar and can be merged together in one term having only one coefficient. Similarly, one can show that when K increases, while the number of coefficients increases exponentially in the dual-input polynomial model, the number of coefficient reduction obtained by using the multi-variable polynomial model also increases considerably. The same concept can be applied if the number of branches of the MIMO transmitter is higher than two. In the following, a mathematical formulation of the multi-variable polynomial model will be given first for a dual-input MIMO system. Then, this model will be generalized for higher number of inputs.

10.2.4.1 Dual-Input Memoryless Model

In the case of a dual-input MIMO transmitter with nonlinear crosstalk, the multi-variable polynomial model in its pass-band format can be expressed for the output of the first branch as follows:

$$\tilde{y}_1(t) = \sum_{k=1}^{K} \sum_{r=0}^{k} h_{kr} [\tilde{x}_1(t)]^{k-r} [\tilde{x}_2(t)]^r \qquad (10.32)$$

This model includes all the cross-terms between $\tilde{x}_1(t)$ and $\tilde{x}_2(t)$ signals that have a nonlinearity order smaller or equal to the maximum linearity order, K. The output of the other branch can be obtained similarly by swapping $\tilde{x}_1(t)$ with the input, $\tilde{x}_2(t)$, of the other branch. In the following analysis of the 2×2 and $N \times N$ MIMO, to avoid redundancy, the analysis will be detailed only for the first branch. Without loss of generality, the index of the output signal is removed.

The baseband version of the multi-variable polynomial model should include only the baseband equivalent of the terms that have a frequency component around the carrier frequency. Therefore, nonlinear terms with even order nonlinearity are not considered in the baseband model and only odd-order nonlinear terms are maintained. The baseband memoryless multi-variable polynomial model is given by:

$$y(n) = \sum_{\substack{k=1 \\ k\ odd}}^{K} \sum_{r=0}^{k} a_{kr} \sum_{k_1=0}^{k-r} \sum_{\substack{k_2=0 \\ k_1+k_2=\frac{k-1}{2}}}^{r} C_{k-r}^{k_1} C_r^{k_2} |x_1(n)|^{2k_1} |x_2(n)|^{2k_2} \cdot [x_1(n)]^{k-r-2k_1} [x_2(n)]^{r-2k_2}$$

$$(10.33)$$

where $a_{kr} = \dfrac{1}{2^{k-1}} h_{kr}$ and $C_m^p = \begin{pmatrix} m \\ p \end{pmatrix} = \dfrac{m!}{(m-p)!p!}$.

The expression in Equation 10.33 can be developed into a matrix form as follows:

$$\mathbf{y} = \mathbf{X}_{\mathbf{x}_1,\mathbf{x}_2} \cdot \mathbf{A} \qquad (10.34)$$

where $\mathbf{y} = [y(1) \quad \cdots \quad y(M)]^T$ is an $M \times 1$ vector representing M samples of the output signal, $\mathbf{A} = [a_{1,0}\ a_{1,1} \quad \cdots \quad a_{k,0} \quad \cdots \quad a_{k,k} \quad \cdots \quad a_{K,0} \quad \cdots \quad a_{K,K}]^T$ is a $P \times 1$ vector of the polynomial coefficients for odd values of k and $P = \sum_{\substack{k=1 \\ k\ odd}}^{K} (k+1)$.

$\mathbf{X}_{\mathbf{x}_1,\mathbf{x}_2} = [\boldsymbol{\beta}_{\mathbf{x}_1,\mathbf{x}_2}^{1,0}\ \boldsymbol{\beta}_{\mathbf{x}_1,\mathbf{x}_2}^{1,1} \quad \cdots \quad \boldsymbol{\beta}_{\mathbf{x}_1,\mathbf{x}_2}^{K,0} \quad \cdots \quad \boldsymbol{\beta}_{\mathbf{x}_1,\mathbf{x}_2}^{K,K}]$ is an $M \times P$ matrix, such that $\boldsymbol{\beta}_{\mathbf{x}_1,\mathbf{x}_2}^{k,r} = \{\beta^{k,r}[x_1(1),x_2(1)] \quad \cdots \quad \beta^{k,r}[x_1(M),x_2(M)]\}^T$ is an $M \times 1$ vector where $\beta^{k,r}$ is defined as:

$$\beta^{k,r}[x_1(n),x_2(n)] = \sum_{k_1=0}^{k-r} \sum_{\substack{k_2=0 \\ k_1+k_2=\frac{k-1}{2}}}^{r} C_{k-r}^{k_1} C_r^{k_2} |x_1(n)|^{2k_1} |x_2(n)|^{2k_2} [x_1(n)]^{k-r-2k_1} [x_2(n)]^{r-2k_2}$$

$$(10.35)$$

with $\mathbf{x}_1 = [x_1(1) \quad x_1(2) \quad \cdots \quad x_1(M)]^T$ and $\mathbf{x}_2 = [x_2(1) \quad x_2(2) \quad \cdots \quad x_2(M)]^T$, which are the $M \times 1$ vectors representing M samples of the baseband input signals.

10.2.4.2 N-Input Memoryless Model

The dual-input multi-variable memoryless polynomial model can be extended to the $N \times N$ MIMO case, where the pass-band representation is given by:

$$\tilde{y} = \sum_{k=1}^{K} \sum_{r_1=0}^{k} \sum_{r_2=0}^{k-r_1} \cdots \sum_{r_{N-1}=0}^{k-(r_1+r_2+\cdots+r_{N-2})} h_{k,r_1,\cdots,r_{N-1}} \tilde{x}_1^{k-(r_1+r_2+\cdots+r_{N-1})} \tilde{x}_2^{r_1} \cdots \tilde{x}_N^{r_{N-1}} \quad (10.36)$$

where $\tilde{x}_i(t)$ and $\tilde{y}(t)$ representing the ith input and the first output pass-band signals, respectively, are replaced by $\tilde{x}_i$ and $\tilde{y}$, respectively, for easiness of representation.

The equivalent baseband model can be shown to have the form:

$$y = \sum_{\substack{k=1 \\ k\,odd}}^{K} \sum_{r_1=0}^{k} \sum_{r_2=0}^{k-r_1} \cdots \sum_{r_{N-1}=0}^{k-(r_1+r_2+\cdots+r_{N-2})} a_{k,r_1,\cdots,r_{N-1}} \sum_{k_1=0}^{k-R} \cdots$$

$$\sum_{\substack{k_N=0 \\ k_1+k_2+\cdots+k_N=\frac{k-1}{2}}}^{r_{N-1}} C_{k-R}^{k_1} \cdots C_{r_{N-1}}^{k_N} |x_1|^{2k_1} \cdots |x_N|^{2k_N} x_1^{k-R-2k_1} x_2^{r_1-2k_2} \cdots x_N^{r_{N-1}-2k_N} \quad (10.37)$$

where $R = r_1 + r_2 + \cdots + r_{N-1}$ and $a_{k,r_1,\cdots,r_{N-1}} = \frac{1}{2^{k-1}} h_{k,r_1,\cdots,r_{N-1}}$.

Here also $x_i(n)$, $(i = 1 \text{ to } N)$, and $y(n)$ representing the ith input and the first output baseband signals, respectively, are replaced by x_i and y, respectively, for easiness of representation.

Equation 10.37 can be expressed in matrix form:

$$y = X_{x_1, \ldots, x_N} \cdot A \quad (10.38)$$

where $y = \left[y(1) \cdots y(M) \right]^T$ is an $M \times 1$ vector representing M samples of the output signal of the first branch, $A = [a_{1,\cdots,0} \cdots a_{1,1,\cdots,0} \cdots\cdots a_{K,0,\cdots,0} \cdots a_{K,K,\cdots,K}]^T$ is a $P \times 1$ vector of the polynomial coefficients for the first branch and P is the total number of coefficients of the N-input memoryless multi-variable polynomial model. $X_{x_1,\cdots,x_N} = [\beta_{x_1,\cdots,x_N}^{1,0,\cdots,0} \; \beta_{x_1,\cdots,x_N}^{1,1,\cdots,0} \; \cdots \; \beta_{x_1,\cdots,x_N}^{K,K,\cdots,0} \; \cdots \; \beta_{x_1,\cdots,x_N}^{K,K,\cdots,K}]$ is an $M \times P$ vector where $\beta_{x_1,\cdots,x_N}^{k,r_1,\cdots,r_{N-1}} = \{\beta^{k,r_1,\cdots,r_{N-1}}[x_1(1),\cdots,x_N(1)] \; \cdots \; \beta^{k,r_1,\cdots,r_{N-1}}[x_1(M),\cdots,x_N(M)]\}^T$ and is an $M \times 1$ vector with $\beta^{k,r_1,\cdots,r_{N-1}}$ is defined as:

$$\beta^{k,r_1,\cdots,r_{N-1}}[x_1, x_2, \cdots, x_N] = \sum_{k_1=0}^{k-R} \cdots \sum_{\substack{k_N=0 \\ k_1+k_2+\cdots+k_N=\frac{k-1}{2}}}^{r_{N-1}} C_{k-R}^{k_1} \cdots C_{r_{N-1}}^{k_N} |x_1|^{2k_1} \cdots |x_N|^{2k_N}$$

$$x_1^{k-R-2k_1} x_2^{r_1-2k_2} \cdots x_N^{r_{N-1}-2k_N} \quad (10.39)$$

In Equation 10.39, the sample indices are removed for easiness and $x_i(m)$ is replaced by x_i.

$x_n = [x_n(1), x_n(2), \cdots, x_n(M)]^T$ is an $M \times 1$ vector representing M samples of the nth input signal.

10.2.4.3　N-Input Memory Model

Similar to the SISO transmitter case, when signals bandwidths in a MIMO transmitter are wide, PAs exhibit memory effects. Consequently, PAs and transmitters modeling and predistortion should consider these memory effects along with the nonlinearity of PAs. Including the memory terms in the N-input multi-variable memoryless polynomial model of Equation 10.37, this model can be extended for the memory case. The N-input memory polynomial model expression is then given by:

$$
y(m) = \sum_{\substack{k=1 \\ k \, odd}}^{K} \sum_{r_1=0}^{k} \cdots \sum_{r_{N-1}=0}^{k-(r_1+\cdots+r_{N-2})} \sum_{q=0}^{Q} a_{k,r_1,\cdots,r_{N-1}}^{q} \sum_{k_1=0}^{k-R} \cdots
$$

$$
\sum_{\substack{k_N=0 \\ k_1+k_2+\cdots+k_N=\frac{k-1}{2}}}^{r_{N-1}} C_{k-R}^{k_1} \cdots C_{r_{N-1}}^{k_N} |x_1(m-q)|^{2k_1} |x_2(m-q)|^{2k_2} \cdots |x_N(m-q)|^{2k_N}
$$

$$
\cdot [x_1(m-q)]^{k-R-2k_1} [x_2(m-q)]^{r_1-2k_2} \cdots [x_N(m-q)]^{r_{N-1}-2k_N} \tag{10.40}
$$

where $x_n(m)$ is the mth sample of the input signal to the nth path of the MIMO system, $y(m)$ is the output signal of the first branch, and Q is the number of delay taps used for the inclusion of the memory effects.

Defining $\mathbf{x}_{nq} = [0_{1 \times q} \, x_n(1) \, \cdots \, x_n(M-q)]^T$ as the $M \times 1$ shifted input vector and, $\mathbf{A} = [a_{1,0,\cdots,0}^{0} \cdots a_{K,K,\cdots,K}^{0} \cdots a_{1,0,\cdots,0}^{Q} \cdots a_{K,K,\cdots,K}^{Q}]^T$ as the $P(Q+1) \times 1$ vector of the polynomial coefficients, and P is the total number of coefficient of the N-input multi-variable memory polynomial model, Equation 10.40 can be rewritten in a matrix format as:

$$
\mathbf{y} = \mathbf{X}_{\mathbf{x}_1,\cdots,\mathbf{x}_N} \cdot \mathbf{A} \tag{10.41}
$$

where

$$
\mathbf{X}_{\mathbf{x}_1,\cdots,\mathbf{x}_M} = \left[\boldsymbol{\beta}_{\mathbf{x}_1,\cdots,\mathbf{x}_N}^{0} \cdots \boldsymbol{\beta}_{\mathbf{x}_1,\cdots,\mathbf{x}_N}^{q} \cdots \boldsymbol{\beta}_{\mathbf{x}_1,\cdots,\mathbf{x}_N}^{Q} \right]
$$

$$
\boldsymbol{\beta}_{\mathbf{x}_1,\cdots,\mathbf{x}_M}^{q} = \left[\beta^{1q} \cdots \beta^{kq} \cdots \beta^{Kq} \right],
$$

$$
\beta^{kq} = \left[\theta_q^{k,0,\cdots,0} \left(x_{1q}, \cdots, x_{nq}, \cdots, x_{Nq} \right) \cdots \theta_q^{k,k,\cdots,k} \left(x_{1q}, \cdots, x_{nq}, \cdots, x_{Nq} \right) \right],
$$

$$
\theta_q^{k,r_1,\cdots,r_{N-1}} = \sum_{k_1=0}^{k-R} \cdots \sum_{k_N=0}^{r_{N-1}} C_{k-R}^{k_1} \cdots C_{r_{N-1}}^{k_N} |x_1(m-q)|^{2k_1} \cdots |x_N(m-q)|^{2k_N}
$$

$$
\begin{array}{c}
k_1+\cdots+k_N=\frac{k-1}{2}
\end{array}
$$

$$
\times [x_1(m-q)]^{k-R-2k_1} [x_2(m-q)]^{r_1-2k_2} \cdots [x_N(m-q)]^{r_{N-1}-2k_N},
$$

and

$$
R = r_1 + r_2 + \cdots + r_{N-1}.
$$

10.3   Modeling and Linearization of Dual-Band Transmitters

Wireless communication systems are rapidly evolving to satisfy the increasing demand for data rates. Multi-band and multi-standard wireless communication systems have caught the attention of recent research activities as a cost effective solution for higher data rates and more optimal use of the frequency spectrum. In such systems, the same RF front end is used for the transmission of two or several wireless communication signals at different carrier frequencies concurrently. While this approach allows better use of the frequency spectrum and better use of the hardware, the concurrent transmission accentuates the distortion problems in the nonlinear components of the transmitter, mainly the PA. Indeed, in addition to the inter-modulation products, cross-band modulation and intra-band modulation products are generated in the multi-band transmission case. Therefore, conventional DPD techniques, such as the Volterra DPD [33–35] or memory polynomial DPD [4, 5] for linearizing each band separately, are not able to fully compensate for the distortions in these multi-band systems.

In order to compensate for these complex distortion terms, one can use conventional DPDs and sample the multi-band signals as a wideband signal. However, this solution is not practical since it will put stringent requirements on the speed of the digital-to-analog converters and analog-to-digital converters. A multi-band DPD technique in which each band's signal is captured and digitized separately, while intending to model all the inter-modulation, intra-modulation, and cross-modulation terms, is then of great importance for the adoption of multi-band transmission architectures. Such objective motivated the recent research activities to propose new DPD architectures suitable for the linearization of multi-band transmitters [36–41]. Two classes of techniques have been proposed. The first class, proposed in [36], consists of a frequency-selective predistortion technique where two independent DPD functions are used in order to alleviate the use of a wideband DPD model, and the nonlinear behavior of the PA is characterized using a large-signal network analyzer. In [41], the frequency selective predistortion technique was extended to include the effects of third and fifth order inter-modulation products. The effectiveness of adding the higher order inter-modulation products on the DPD linearization is investigated for narrow-spaced input signals in dual-band and tri-band cases. In the second class of multi-band DPD systems, dynamic characterization using the actual signals that are sent to the PA is adopted, and a multi-input DPD architecture that takes as the input the baseband equivalent of the signals in each band, are proposed. The first attempt to model and linearize multi-band transmitters using a dynamic characterization was provided in [37, 38]. In these works, a multi-dimensional memory polynomial DPD is proposed to linearize a widely spaced dual-band PA transmitter. Additional modeling and reverse modeling using multi-input DPD architectures based on Volterra series [42–44] had been proposed to improve the linearization performance and to reduce the model complexity. This section will focus on the latter class of multi-band predistortion systems as they are more suitable for implementation in field deployed systems due to the nature of the signals used in the characterization step of frequency selective multi-band DPD systems.

10.3.1 Generalization of the Polynomial Model to the Dual-Band Case

The nonlinearity in concurrent dual-band PAs results in more complex distortions than in the single band case due to the cross-modulation products between the two bands, which results in in-band and out-of-band inter-modulation products [38, 39].

Figure 10.10 shows the power spectral densities (PSDs) at the input and output of a concurrent dual-band PA. The input signals for the upper and lower bands were two-tone signals with different spacing. In the output signal, one can distinguish two different parts of the spectrum:

- *The in-band inter-modulation and cross-modulation products:* These products fall within the useful bands of each of the two signals and therefore should be present in the low-pass equivalent model of the PA.
- *The out-of-band inter-modulation products:* These products fall outside the useful bands and therefore can be ignored.

The distortion products generated by each band are shown with continuous arrows having the same width used to represent the signal in that band. For examples, the continuous thin arrows shown in Figure 10.10 show the signal around the angular frequency ω_1 and its contribution to the distortion products in-band. This contribution includes the in-band inter-modulation products that fall around the carrier angular frequency ω_1 and the cross-modulation products that fall around the carrier angular frequency ω_2 of the second band. This illustration demonstrates that both signals contribute to the low-pass equivalent of the PA nonlinear behavior around each of the carrier frequencies. Therefore, even if modeled or linearized separately, each band behavior has to be a function of the signals present in the two bands. A more complex multi-input model is therefore required for the characterization and linearization of concurrent dual-band transmitters.

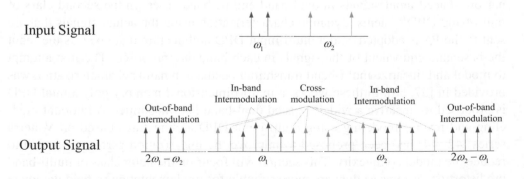

Figure 10.10   Effect of power amplifier nonlinearity on concurrent dual-band transmission

One approach consists of having the dual-band model/DPD composed of two cells, one for each communication band. Each cell has a dual-input that consists of the signals to be transmitted in each of the two bands. This guarantees that the characterization of each band includes the in-band inter-modulation and cross-modulation products. In order to better analyze the behavior of dual-band non-linear distortion in PAs, we are considering the case of approximating the PA behavior to a general memoryless pass-band nonlinear model, which was given by Equation 10.27.

In case of a multi-band transmission, the input of the pass-band model is given by:

$$\tilde{x}(t) = \sum_{b=1}^{B} \mathrm{Re}[x_b(t)e^{j\omega_b t}] \tag{10.42}$$

where ω_b is the angular carrier frequency, $x_b(t)$ is the complex envelope of the band-pass signal, $\tilde{x}_b(t) = \mathrm{Re}[x_b(t)e^{j\omega_b t}]$, present in the bth band; and B is the total number of bands in the concurrent multi-band transmission.

In the following analysis, a dual-band transmission is considered. This analysis can be extended to any number of bands. In the case of a dual-band transmission, the pass-band signal at the input of the transmitter is given by:

$$\tilde{x}(t) = \tilde{x}_1(t) + \tilde{x}_2(t) = \sum_{b=1}^{2} \mathrm{Re}[x_b(t)e^{j\omega_b t}]. \tag{10.43}$$

The pass-band signal at the output of the concurrent dual-band PA, $\tilde{y}(t)$, can be calculated by plugging the expression of the input signal of Equation 10.43 into the expression of the output given in Equation 10.27. After manipulation, the expression of the output signal, $\tilde{y}(t)$, can be derived as shown in Equation 10.44. Details of the calculation can be obtained in [43, 45]. Here for easiness of notation, the time argument (t) is removed:

$$
\begin{aligned}
\tilde{y} = \sum_{k=0}^{K-1} \sum_{r=0}^{k} \sum_{k_1=0}^{k-r} \sum_{k_2=0}^{r} \frac{1}{2^k} a_{kr} C_k^r C_{k-r}^{k_1} C_r^{k_2} & \cdot [(x_1|x_1|^{2k_1}|x_2|^{2k_2}) \cdot e^{j\omega_1 t} \\
& + (x_2|x_1|^{2k_1}|x_2|^{2k_2}) \cdot e^{j\omega_2 t} \\
& + (x_1^* x_2^2 |x_1|^{2(k_1-1)}|x_2|^{2k_2}) \cdot e^{j(2\omega_2-\omega_1)t} \\
& + (x_1^2 x_2^* |x_1|^{2k_1}|x_2|^{2(k_2-1)}) \cdot e^{j(2\omega_1-\omega_2)t} \\
& + (x_1^{*2} x_2^3 |x_1|^{2(k_1-2)}|x_2|^{2k_2}) \cdot e^{j(3\omega_2-2\omega_1)t} + \cdots]
\end{aligned} \tag{10.44}
$$

where $C_n^k = \frac{n!}{k!(n-k)!}$ is the binomial coefficient n choose k.

This general equation has components around the two carrier frequencies ω_1 and ω_2 along with other higher-order inter-modulation products.

If the case of concurrent dual-band transmission, the frequency spacing between the two carrier frequencies ω_1 and ω_2 can be considered large enough, so that the inter-band third-order inter-modulation products around $(2\omega_2 - \omega_1)$ and $(2\omega_1 - \omega_2)$ and further higher order inter-band inter-modulation products can be considered out-of-band and therefore can be filtered out. In the baseband representation, these terms are removed and only the in-band inter-modulation and cross-modulation products are kept. The complex envelope of the output signal around each of the two carriers is then given by:

$$y_1 = \sum_{k=0}^{K-1} \sum_{r=0}^{k} \sum_{k_1=0}^{k-r} \sum_{k_2=0}^{r} \frac{1}{2^{k-1}} a_{kr}^{(1)} C_k^r C_{k-r}^{k_1} \; C_r^{k_2} \cdot (x_1 |x_1|^{2k_1} |x_2|^{2k_2}) \tag{10.45}$$

$$y_2 = \sum_{k=0}^{K-1} \sum_{r=0}^{k} \sum_{k_1=0}^{k-r} \sum_{k_2=0}^{r} \frac{1}{2^{k-1}} a_{kr}^{(2)} C_k^r C_{k-r}^{k_1} \; C_r^{k_2} \cdot (x_2 |x_2|^{2k_1} |x_1|^{2k_2}) \tag{10.46}$$

It is important to note that, theoretically, the coefficients $a_{kr}^{(1)}$ and $a_{kr}^{(2)}$ are identical and are equal to a_{kr}. However, because the PA behavior is different in each band, in practice, these coefficients are different and for this reason, different notations are used here. It is also important to note that the output signal in each band is a function of the two input signals and, more precisely, includes cross-terms with different nonlinearity orders while maintaining the same phase as the complex envelope of that corresponding band. Finally, it is important to mention that while for behavioral modeling only odd order nonlinearity terms are included, it was shown that in the case of DPD, including even-order terms can improve the linearization performance [17]. In general, Equations 10.45 and 10.46 can be expressed in their general form including even and odd terms and rearranging all the terms with the same form together as follows:

$$y_1 = \sum_{k=0}^{K-1} \sum_{r=0}^{k} A_{kr}^{(1)} \cdot (x_1 |x_1|^{k-r} |x_2|^{r}) \tag{10.47}$$

$$y_2 = \sum_{k=0}^{K-1} \sum_{r=0}^{k} A_{kr}^{(2)} \cdot (x_2 |x_2|^{k-r} |x_1|^{r}) \tag{10.48}$$

where $A_{kr}^{(1)}$ and $A_{kr}^{(2)}$ are the model coefficients, which are function of sets of $a_{kr}^{(1)}$ and $a_{kr}^{(2)}$, respectively.

10.3.2   Two-Dimensional (2-D) Memory Polynomial Model for Dual-Band Transmitters

The memoryless pass-band model can be further generalized to include the memory effects exhibited by the dual-band PA. Starting from a memory polynomial pass-band model given by:

$$\tilde{y}(n) = \sum_{m=0}^{M} \sum_{k=1}^{K} h_{k,m} \tilde{x}^k(n-m)$$ (10.49)

where N and M are the nonlinearity order and the memory order of the model, respectively. Using similar analysis as in the memoryless case, a 2-D MP baseband equivalent model can be developed for behavioral modeling or predistortion of the dual-band transmitter [37]. Similar to the memoryless dual-band model, this memory model is composed of two cells, one for each band. The model estimated outputs are given by the following equations:

$$y_1(n) = \sum_{k=0}^{K-1} \sum_{r=0}^{k} \sum_{m=0}^{M} A_{k,r,m}^{(1)} x_1(n-m)|x_1(n-m)|^{k-r}|x_2(n-m)|^r$$ (10.50)

$$y_2(n) = \sum_{k=0}^{K-1} \sum_{r=0}^{k} \sum_{m=0}^{M} A_{k,r,m}^{(2)} x_2(n-m)|x_2(n-m)|^{k-r}|x_1(n-m)|^r$$ (10.51)

where $y_1(n)$ and $y_2(n)$ are the complex envelopes, estimated by the baseband 2-D memory polynomial model, of the output signals in the two bands centered around the two fundamental frequencies ω_1 and ω_2, respectively; $x_1(n)$ and $x_2(n)$ are baseband complex envelopes of the input signals in the two bands; and $A_{k,r,m}^{(1)}$ and $A_{k,r,m}^{(2)}$ are the model coefficients for the two bands.

Similar to the memoryless case, in the case of the 2-D memory polynomial model, the output signal in each band includes cross-terms with different nonlinearity orders while maintaining the same phase as the complex envelope of that corresponding band. Therefore, it has no information of the phase of the input signal in the other band. Consequently, for a given band, the model cannot account for the variations and effects attributed to the phase of the other band. In order to account for such effects, a Volterra based multi-band model is needed.

10.3.3   Phase-Aligned Multi-Band Volterra DPD

A truncated Volterra series can be used to model the nonlinearity and memory effects in a single-band PA as was described in Chapter 2. The expression of the pass-band Volterra series model limited to the third order nonlinearity and a finite memory depth is given by:

$$\tilde{y}(n) = \sum_{m=0}^{M_1} a_m^{(1)} \tilde{x}(n-m) + \sum_{l=0}^{M_2} \sum_{m=0}^{M_1} a_{l,m}^{(2)} \tilde{x}(n-l)\tilde{x}(n-m)$$

$$+ \sum_{p=0}^{M_3} \sum_{l=0}^{M_2} \sum_{m=0}^{M_1} a_{p,l,m}^{(3)} \tilde{x}(n-p)\tilde{x}(n-l)\tilde{x}(n-m)$$ (10.52)

where $\tilde{x}(n)$ and $\tilde{y}(n)$ are the pass-band input and output signals, respectively; $a^{(i)}_{\ldots}$ are the Volterra kernels for nonlinearity of order i and different delay lengths; and M_j are the memory depths of the model. Here, the continuous time representation for the pass-band model that was used in Chapter 2 is replaced by a discrete time representation in order to illustrate more clearly the memory effects observed in the baseband model. This discrete time representation of the pass-band model will be used in the remaining of this section.

This Volterra series model is complex since its number of coefficients increases exponentially with the degree of nonlinearity and memory depth of the system. Therefore, this model cannot be used for systems with high order nonlinearity or memory depths as very high number of coefficients are not practical to extract. To reduce the complexity issue in the Volterra series model, different variants of this model were proposed [1, 2, 35, 46–48]. In the following analysis, one complexity-reduced variant, the simplified Volterra model, is used. This model is given by:

$$\tilde{y}(n) = \tilde{y}_{na}(n) + \tilde{y}_a(n) \tag{10.53}$$

where $\tilde{y}_{na}(n)$ is the part of the output signal that includes the nonlinearity terms with no phase alignment, and $\tilde{y}_a(n)$ is the part of the output signal that includes the nonlinearity terms with phase alignment.

By separating the non-aligned and aligned terms, this variant offers more freedom in modeling the mildly nonlinear phase-aligned terms with a low-order nonlinearity and sufficiently high order memory depth, while keeping acceptable high order nonlinearity for modeling the non-aligned term. As a result, lesser number of coefficients is needed and better modeling accuracy is achieved. Moreover, separate identification of the non-aligned part's coefficients and the aligned part's coefficients reduces the complexity of the identification procedure [2, 46].

The non-aligned function $\tilde{y}_{na}(n)$ can be expressed by a power series as follows:

$$\tilde{y}_{na}(n) = \sum_{m=0}^{M_n} \sum_{k=1}^{K_n} a_m^{(k)} \tilde{x}^k(n-m) \tag{10.54}$$

where $a_m^{(k)}$, M_n, and K_n are the coefficients, the memory order and the nonlinearity order of the pass-band non-aligned function. The mildly nonlinear aligned function $\tilde{y}_a(n)$ can be expressed by a dynamic Volterra filter, where only the bi-dimensional second-order cross-terms (i.e., the products with two different time delays and with arbitrary order of nonlinearities) are kept [48]. Its expression is given by:

$$\tilde{y}_a(n) = \sum_{k=1}^{K_a} \sum_{r=0}^{k-1} \sum_{m=0}^{M_1-1} \sum_{l=m+1}^{M_2-1} a_{l,m}^{(k,r)} \tilde{x}^r(n-l) \tilde{x}^{k-r}(n-m) \tag{10.55}$$

where M_1 and M_2 are the memory depths of the second-order cross-terms, K_a is the nonlinearity order, and $a_{l,m}^{(k,r)}$ are the coefficients of the bandpass phase-aligned function.

Therefore, the total output of the simplified Volterra model in the single-band case is given by:

$$\tilde{y}(n) = \sum_{m=0}^{M_n}\sum_{k=1}^{K_n} a_m^{(k)}\tilde{x}^k(n-m) + \sum_{k=1}^{K_a}\sum_{r=0}^{k-1}\sum_{m=0}^{M_1-1}\sum_{l=m+1}^{M_2-1} a_{l,m}^{(k,r)}\tilde{x}^r(n-l)\tilde{x}^{k-r}(n-m) \quad (10.56)$$

The non-aligned component $y_{na}(n)$ of the baseband output signal is a memory polynomial function of the input signal. Its generalization to the dual-band case is similar to the dual-band memory polynomial model given in Equations 10.50 and 10.51. The generalization of this non-aligned component $y_{na}(n)$ to the dual-band case is given by:

$$y_{na,1}(n) = \sum_{m=0}^{M_n}\sum_{k=0}^{K_n-1}\sum_{r=0}^{k} A_{m,k,r}^{(1)}x_1(n-m)|x_1(n-m)|^{k-r}|x_2(n-m)|^r \quad (10.57)$$

$$y_{na,2}(n) = \sum_{m=0}^{M_n}\sum_{k=0}^{K_n-1}\sum_{r=0}^{k} A_{m,k,r}^{(2)}x_2(n-m)|x_2(n-m)|^{k-r}|x_1(n-m)|^r \quad (10.58)$$

where $A_{m,k,r}^{(i)}$ is the set of baseband coefficients of the non-aligned part $y_{na,i}(n)$ around angular frequency ω_i, and $i \in \{1,2\}$.

For the aligned part, substituting the dual-band input expression of $\tilde{x}(n)$ given by Equation 10.43 in Equation 10.55, the aligned function $\tilde{y}_a(n)$ will be given as:

$$\tilde{y}_a(n) = \sum_{k=1}^{K_a}\sum_{r=0}^{k-1}\sum_{m=0}^{M_1-1}\sum_{l=m+1}^{M_2-1} a_{l,m}^{(k,r)}\cdot[\tilde{x}_1(n-m)+\tilde{x}_2(n-m)]^{k-r}\cdot[\tilde{x}_1(n-l)+\tilde{x}_2(n-l)]^r$$
$$(10.59)$$

To derive the pass-band model from the baseband equivalent model, only terms located around the two fundamental frequencies ω_1 and ω_2 in the expression of the pass-band aligned model are considered. Other terms can be filtered out. The baseband equivalent of the aligned function of Equation 10.59 can be obtained after manipulation as shown in Equation 10.60 below. Details of the calculation for more simplified cases can be found in [45, 49].

$$y_{a,i}(n)$$

$$
= \sum_{m=0}^{M_1-1} \sum_{l=m+1}^{M_2-1} \left\{ \psi_{0,l,m}^{(i)}[x_1(n), x_2(n)] \cdot \sum_{\substack{r_1+r_2+r_3+r_4=1 \\ r_3+r_4 \neq 0}}^{K_a-1} A_{0,l,m}^{(i,r_1,r_2,r_3,r_4)} \cdot F^{(m,l,r_1,r_2,r_3,r_4)}[x_1(n), x_2(n)] \right\}
$$

$$
+ \sum_{m=0}^{M_1-1} \sum_{l=m+1}^{M_2-1} \left\{ \psi_{1,l,m}^{(i)}[x_1(n), x_2(n)] \cdot \sum_{\substack{r_1+r_2+r_3+r_4=0}}^{K_a-1} A_{1,l,m}^{(i,r_1,r_2,r_3,r_4)} \cdot F^{(m,l,r_1,r_2,r_3,r_4)}[x_1(n), x_2(n)] \right\}
$$

$$
+ \sum_{j=2}^{7} \sum_{m=0}^{M_1-1} \sum_{l=m+1}^{M_2-1} \left\{ \psi_{j,l,m}^{(i)}[x_1(n), x_2(n)] \cdot \sum_{\substack{r_1+r_2+r_3+r_4=0}}^{K_a-3} A_{j,l,m}^{(i,r_1,r_2,r_3,r_4)} \cdot F^{(m,l,r_1,r_2,r_3,r_4)}[x_1(n), x_2(n)] \right\}
$$

$$(10.60)$$

where $A_{j,l,m}^{(i,r_1,r_2,r_3,r_4)}$ are the set of baseband coefficients of the aligned part $y_{a,i}(n)$ around angular frequency ω_i, and $i \in \{1,2\}$.

$F^{(m,l,r_1,r_2,r_3,r_4)}[x_1(n), x_2(n)]$ is the compound envelope function given by:

$$
F^{(m,l,r_1,r_2,r_3,r_4)}[x_1(n), x_2(n)] = |x_1(n-m)|^{r_1} |x_2(n-m)|^{r_2} |x_1(n-l)|^{r_3} |x_2(n-l)|^{r_4}
$$

$$(10.61)$$

and $\psi_{j,l,m}^{(i)}[x_1(n), x_2(n)]$ are the phase alignment functions given for $i = 1$ by:

$$
\psi_{0,l,m}^{(1)}[x_1(n), x_2(n)] = x_1(n-m)
$$

$$
\psi_{1,l,m}^{(1)}[x_1(n), x_2(n)] = x_1(n-l)
$$

$$
\psi_{2,l,m}^{(1)}[x_1(n), x_2(n)] = x_1^2(n-m)x_1^*(n-l)
$$

$$
\psi_{3,l,m}^{(1)}[x_1(n), x_2(n)] = x_1(n-m)x_2^*(n-l)x_2(n-m) \qquad (10.62)
$$

$$
\psi_{4,l,m}^{(1)}[x_1(n), x_2(n)] = x_1(n-m)x_2^*(n-m)x_2(n-l)
$$

$$
\psi_{5,l,m}^{(1)}[x_1(n), x_2(n)] = x_1^2(n-l)x_1^*(n-m)
$$

$$
\psi_{6,l,m}^{(1)}[x_1(n), x_2(n)] = x_1(n-l)x_2^*(n-l)x_2(n-m)
$$

$$
\psi_{7,l,m}^{(1)}[x_1(n), x_2(n)] = x_1(n-l)x_2^*(n-m)x_2(n-l)
$$

Similar expression can be obtained for $i = 2$ as follows:

$$\psi_{0,l,m}^{(2)}[x_1(n), x_2(n)] = x_2(n - m)$$

$$\psi_{1,l,m}^{(2)}[x_1(n), x_2(n)] = x_2(n - l)$$

$$\psi_{2,l,m}^{(2)}[x_1(n), x_2(n)] = x_2^2(n - m)x_2^*(n - l)$$

$$\psi_{3,l,m}^{(2)}[x_1(n), x_2(n)] = x_2(n - m)x_1^*(n - l)x_1(n - m) \qquad (10.63)$$

$$\psi_{4,l,m}^{(2)}[x_1(n), x_2(n)] = x_2(n - m)x_1^*(n \quad m)x_1(n - l)$$

$$\psi_{5,l,m}^{(2)}[x_1(n), x_2(n)] = x_2^2(n - l)x_2^*(n - m)$$

$$\psi_{6,l,m}^{(2)}[x_1(n), x_2(n)] = x_2(n - l)x_1^*(n - l)x_1(n - m)$$

$$\psi_{7,l,m}^{(2)}[x_1(n), x_2(n)] = x_2(n - l)x_1^*(n - m)x_1(n - l)$$

Similarly to the memoryless and memory polynomial based models, the pass-band Volterra model has no even-order terms. However, it was shown that even-order terms can be added to the expression of the baseband equivalent model in order to enhance the basis set, thus reducing the modeling error especially in the DPD case [17].

Therefore, the baseband equivalent output waveforms of the phase-aligned dual-band Volterra model are obtained from Equations 10.57, 10.58, and 10.60 as follows:

$$y_1(n) = y_{na,1}(n) + y_{a,1}(n) \qquad (10.64)$$

$$y_2(n) = y_{na,2}(n) + y_{a,2}(n) \qquad (10.65)$$

where $y_1(n)$ and $y_2(n)$ are the baseband equivalent waveforms of the signals at the output of the dual-band power amplifier estimated by the phase-aligned dual-band Volterra model around the fundamental frequencies ω_1 and ω_2, respectively.

10.4   Application of MIMO and Dual-Band Models in Digital Predistortion

In order to provide an idea on the expected linearization performance of the different models discussed in this chapter, this sub-section presents some typical measurement results of linearized PAs first in the case of a MIMO system with nonlinear crosstalk then in the case of concurrent dual-band transmission.

10.4.1   Linearization of MIMO Systems with Nonlinear Crosstalk

The measurement results showing the performance of the crossover MIMO DPD in the presence of crosstalk are given in the presence of nonlinear crosstalk. First, the effect of crosstalk on the quality of linearization using conventional memory polynomial is shown in Figure 10.11. This figure shows the spectrum of the output signal from one PA branch of a MIMO system in different scenarios:

1. No linearization is applied to reduce the PA nonlinear distortion,
2. Conventional memory polynomial predistortion is applied for MIMO system in the presence of:
   a. No crosstalk;
   b. −40 dB nonlinear crosstalk;
   c. −30 dB nonlinear crosstalk;
   d. −20 dB nonlinear crosstalk.

From this figure, one can conclude that conventional memory polynomial performs well only in the absence of crosstalk. The higher the crosstalk value, the worse is the performance of the predistortion and therefore the higher is the residual distortion observed at the output of the linearized amplifier.

The performance of the crossover MIMO predistortion is analyzed in Figure 10.12 and Table 10.1 and compared to conventional memory polynomial linearization for a MIMO transmitter having −20 dB nonlinear crosstalk. Figure 10.12 shows the plots

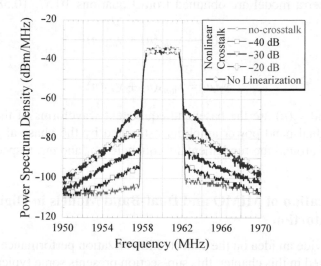

Figure 10.11   Performance of conventional single-input single-output linearization in the presence of nonlinear crosstalk in MIMO systems [21]. ©2009 IEEE. Reprinted, with permission, from S. A. Bassam *et al.*, "Crossover digital predistorter for the compensation of crosstalk and nonlinearity in MIMO transmitters," *IEEE Transactions on Microwave Theory and Techniques*, May 2009

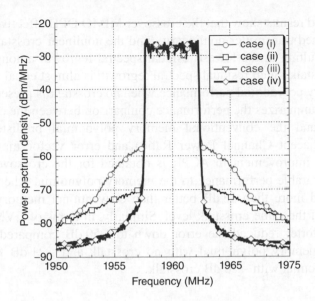

Figure 10.12 Performance of MIMO DPD in the presence of nonlinear crosstalk in MIMO systems [21]. ©2009 IEEE. Reprinted, with permission, from S. A. Bassam *et al.*, "Crossover digital predistorter for the compensation of crosstalk and nonlinearity in MIMO transmitters," *IEEE Transactions on Microwave Theory and Techniques*, May 2009

Table 10.1 EVM and ACPR performance of MIMO DPD compared to conventional single-band linearization in the presence of nonlinear crosstalk in MIMO systems [21]

Scenario	EVM (dB)	ACPR in 5 MHz offset (dBc)
Without DPD and −20 dB nonlinear crosstalk	−20.15	40.66
Conventional DPD and −20 dB nonlinear crosstalk	−21.22	43.31
Crossover MIMO DPD and −20 dB nonlinear crosstalk	−49.71	56.81
Conventional DPD and without nonlinear crosstalk	−53.69	58.23

of the power spectrum of a wideband code division multiple access (WCDMA) signal at the output of the MIMO transmitter for the following four cases:

- no DPD and −20 dB crosstalk;
- conventional memory polynomial and −20 dB crosstalk;
- crossover MIMO DPD and −20 dB crosstalk; and
- conventional memory polynomial predistortion and no crosstalk.

The measured results show that the crossover MIMO DPD effectively compensates for the combined transmitter nonlinearity and the nonlinear crosstalk in the MIMO transmitter resulting in the lowest residual spectral regrowth among the cases with nonlinear crosstalk. Such residual spectral regrowth is almost equal to the one of the case of memory polynomial performance when no crosstalk is present.

Table 10.1 summarizes the performance comparison between the crossover MIMO predistortion and the conventional memory polynomial predistortion in terms of ACPR (Adjacent Channel Power Ratio) and error vector magnitude (EVM). Around 16 dB improvement in ACPR is obtained for the crossover MIMO DPD, which is comparable performance to the memory polynomial when no crosstalk is considered and more than 13 dB better than conventional memory polynomial in the presence of the same crosstalk level. Similarly, in terms of EVM, the crossover MIMO predistorter reduced the error down to −50 dB compared to −53 dB for conventional memory polynomial with no crosstalk and −21 dB for conventional memory polynomial with −20 dB crosstalk.

10.4.2   Linearization of Concurrent Dual-Band Transmitters Using a 2-D Memory Polynomial Model

Example of validation of the performance of the 2-D memory polynomial model in the linearization of concurrent dual-band transmission is provided in this section. A dual-band signal composed of an orthogonal frequency division multiplexing (OFDM) signal of bandwidth 10 MHz centered around a carrier frequency equal to 1.9 GHz and a WCDMA signal of 5 MHz bandwidth centered around a carrier frequency equals to 2 GHz is used for the validation. This dual-band signal is predistorted using the 2-D memory polynomial model and then sent through a dual-band PA. The results are summarized in Figure 10.13 and Table 10.2. Figure 10.13 shows the frequency spectra of the nonlinearized and linearized signals in each of the two bands. The 2-D memory polynomial DPD was able to reduce the spectral regrowth caused by the PA nonlinearity around each of the two bands. More than 10 dB reduction is observed in the first band with the OFDM signal while almost 20 dB improvement is observed in the second band with the WCDMA signal. Table 10.2 summarizes the performance of the linearization using 2-D memory polynomial in terms of residual normalized mean squared error (NMSE) and in terms of ACPR. Compared to the case where no predistortion is used, the NMSE was improved when using the 2-D memory polynomial by more than 10 dB in the lower band and more than 15 dB in the second band, while the ACPR was improved by more than 10 dB in the lower band and by almost 15 dB in the upper band to reach about −55 dBc in both cases. These results show that the 2-D memory polynomial model, when used as a predistorter model, is able to correct for most of the in-band inter-modulation and cross-modulation products generated by the PA when driven by a dual-band signal.

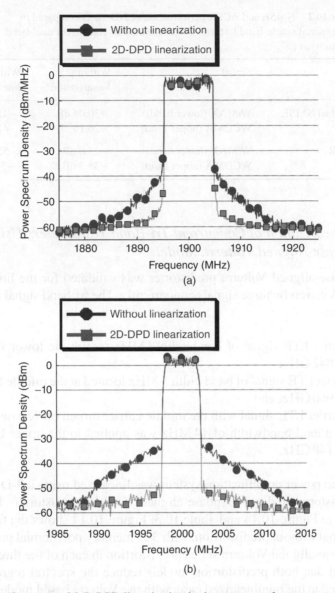

Figure 10.13 Performance of 2-D-DPD linearization in the presence of concurrent dual-band transmission (a) band I with a 10 MHz OFDM signal (©2009 IEEE. Reprinted, with permission, from S. A. Bassam *et al.*, "2-D digital predistortion (2-D-DPD) architecture for concurrent dual-band transmitters," *IEEE Transactions on Microwave Theory and Techniques*, Oct. 2011) and (b) band II with a 5 MHz WCDMA signal (©2009 IEEE. Reprinted, with permission, from S. A. Bassam *et al.*, "2-D digital predistortion (2-D-DPD) architecture for concurrent dual-band transmitters," *IEEE Transactions on Microwave Theory and Techniques*, Oct. 2011) [21, 37].

Table 10.2 NMSE and ACPR performance of 2-D-DPD compared to conventional single-band linearization in the case of concurrent dual-band transmission [37]

	Signal	Without linearization	With linearization
Residual NMSE	WiMAX (lower band)	−30.04 dB	−41.05 dB
	WCDMA (upper band)	−26.19 dB	−42.51 dB
ACPR	WiMAX (lower band)	−45.0 dBc	−55.8 dBc
	WCDMA (upper band)	−38.3 dBc	−54.6 dB

10.4.3 *Linearization of Concurrent Tri-Band Transmitters Using 3-D Phase-Aligned Volterra Model*

The 3-D phase-aligned Volterra predistorter was validated for the linearization of a wideband PA driven by three signals concurrently. The tri-band signal is composed of the following:

1. a two-carrier LTE signal of bandwidth 8 MHz sent at the lower frequency band around 1.842 GHz;
2. a one-carrier LTE signal of bandwidth 5 MHz located at the middle frequency band around 1.960 GHz; and
3. a three-carrier LTE signal with the middle carrier turned off (carrier configuration 101) and a total bandwidth of 14 MHz was applied in the upper frequency band around 2.140 GHz.

The tri-band power amplification system was linearized using a 3-D memory polynomial predistorter and a 3-D phase-aligned Volterra predistorter. The results are summarized in Figure 10.14 and Table 10.3. Figure 10.14 shows the frequency spectra of the signal without predistortion, with 3-D memory polynomial predistortion and with 3-D phase-aligned Volterra series predistortion in each of the three bands. It can be concluded that both predistortion models reduce the spectral regrowth considerably compared to the nonlinearized case, with the Volterra based model offering extra few decibels reduction since it includes additional cross-terms that are not present in the 3-D memory polynomial. The ACPR performances are summarized in Table 10.3. These results confirm that the 3-D phase-aligned Volterra model performs slightly better that the 3-D memory polynomial model in the linearization of the tri-band system. The obtained ACPR values using the Volterra based model are around −45 dBc for the lower band, −50 dBc for the middle band, and −55 dBc for the upper band in the case of the 3-D phase-aligned Volterra series model. The two models were used with approximately the same number of coefficients; 105 coefficients for the memory polynomial based model versus 126 coefficients for the Volterra based model.

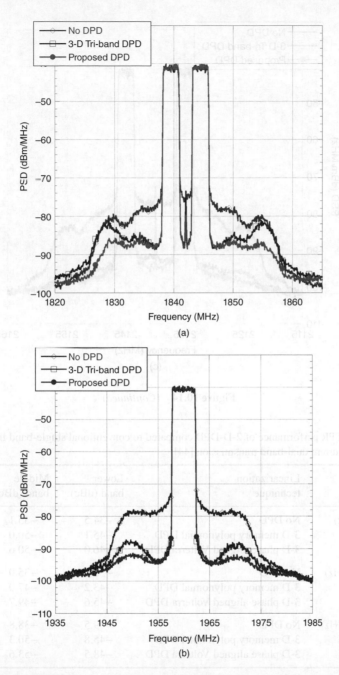

Figure 10.14 Performance of 3-D phase-aligned Volterra DPD linearization in the presence of concurrent tri-band-band transmission and comparison with the 3-D memory polynomial DPD (a) band I; (b) band II; and (c) band III [44]. ©2009 IEEE. Reprinted, with permission, from M. Younes *et al.*, "Linearization of concurrent tri-band transmitters using 3-D phase-aligned pruned Volterra model," *IEEE Transactions on Microwave Theory and Techniques*, Dec. 2013

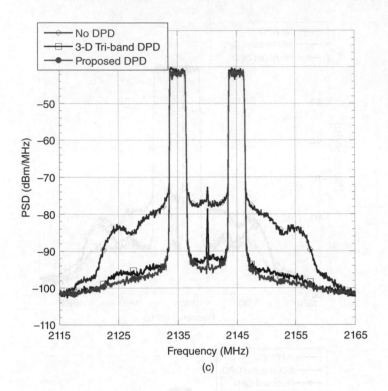

Figure 10.14 (*Continued*)

Table 10.3 ACPR performance of 2-D-DPD compared to conventional single-band linearization in the case of concurrent dual-band transmission [44]

	Linearization technique	Lower band (dBc)	Middle band (dBc)	Upper band (dBc)
ACPR I (5 MHz)	No DPD	−34.5	−35.1	−37.8
	3-D memory polynomial DPD	−45.1	−50.0	−55.0
	3-D phase aligned Volterra DPD	−46.0	−50.6	−56.5
ACPR II (10 MHz)	No DPD	−35.6	−35.0	−39.5
	3-D memory polynomial DPD	−43.2	−47.0	−55.1
	3-D phase aligned Volterra DPD	−45.6	−49.7	−57
ACPR III (15 MHz)	No DPD	−39.5	−38.8	−43.6
	3-D memory polynomial DPD	−45.8	−50.3	−56.1
	3-D phase aligned Volterra DPD	−48.5	−53.6	−57.5

References

[1] Zhu, A. and Brazil, T.J. (2004) Behavioral modeling of RF power amplifiers based on pruned Volterra series. *IEEE Microwave Wireless Component Letters*, **14** (12), 563–565.
[2] Zhu, A., Pedro, J.C. and Brazil, T.J. (2006) Dynamic deviation reduction-based Volterra behavioral

modeling of RF power amplifiers. *IEEE Transactions on Microwave Theory and Techniques*, **54** (12), 4323–4332.

[3] Kim, J. and Konstantino, K. (2001) Digital predistortion of wideband signals based on power ampliflier model with memory. *IEEE Electronic Letters*, **37** (23), 1417–1418.

[4] Raich, R., Qian, H. and Zhou, G.T. (2004) Orthogonal polynomials for power amplifier modeling and predistorter design. *IEEE Transactions on Vehicular Technologies*, **53** (5), 1468–1479.

[5] Helaoui, M., Boumaiza, S., Ghazel, A. and Ghannouchi, F.M. (2006) Power and efficiency enhancement of 3G multicarrier amplifiers using digital signal processing with experimental validation. *IEEE Transactions on Microwave Theory and Techniques*, **54** (4), 1396–1404.

[6] A. G. K. C. Lim, V. Sreeram, and G. Wang, A new technique for digital compensation in IQ modulator. Proceedings IEEE Asia-Pacific Conference on Circuits and Systems, Tainan, Taiwan, December 2004, pp. 493–496, 2004.

[7] A. Cantoni, and J. Tuthill, Digital compensation of frequency dependent imperfections in direct conversion I-Q modulators. Digest IEEE International Symposium on Circuits and Systems, New Orleans, LA, May 2007, pp. 269–272, 2007.

[8] Ebadi, Z.S. and Saleh, R. (2008) Adaptive compensation of RF front-end nonidealities in direct conversion receivers. *IEEE Transactions on Circuits and Systems II: Express Briefs*, **55** (4), 354–358.

[9] Sun, M., Yu, J. and Hsu, T. (2008) Estimation of carrier frequency offset with I/Q mismatch using pseudo-offset injection in OFDM systems. *IEEE Transactions on Circuits and Systems I: Regular Papers*, **55** (3), 943–952.

[10] Ding, L., Ma, Z., Morgan, D.R. *et al.* (2008) Compensation of frequency-dependent gain/phase imbalance in predistortion linearization systems. *IEEE Transactions on Circuits and Systems I: Regular Papers*, **55** (1), 390–397.

[11] Antilla, L., Handel, P. and Valkama, M. (2010) Joint mitigation of power amplifier and I/Q modulator impairments in broadband direct-conversion transmitters. *IEEE Transactions on Microwave Theory and Techniques*, **58** (4), 730–739.

[12] D. Saffar, N. Boulejfen, F. M. Ghannouchi, and M. Helaoui, A joint hardware impairment and distortion compensation of wireless MIMO transmitters. Proceedings IEEE International New Circuits and Systems Conference, Paris, France, June 2013, pp. 1–4, 2013.

[13] Saffar, D., Boulejfen, N., Ghannouchi, F.M. *et al.* (2013) A compound structure and a single-step identification procedure for I/Q and DC offset impairments and nonlinear distortion modeling and compensation in wireless transmitters. *International Journal of RF and Microwave Computer-Aided Engineering*, **23** (3), 367–377.

[14] Rawat, K., Rawat, M. and Ghannouchi, F.M. (2010) Compensating I–Q imperfections in hybrid RF/Digital predistortion with an adapted lookup table implemented in an FPGA. *IEEE Transactions on Circuits and Systems II: Express Briefs*, **57** (5), 389–393.

[15] Gharaibeh, K. (2011) *Nonlinear Distortion in Wireless Systems*, John Wiley & Sons, Inc, Hoboken, NJ.

[16] Ermolova, N.Y. (2001) Spectral analysis of nonlinear amplifier based on the complex gain Taylor series expansion. *IEEE Communication Letters*, **5** (12), 465–467.

[17] Ding, L. and Zhou, G.T. (2004) Effects of even-order nonlinear terms on power amplifier modeling and predistortion linearization. *IEEE Transactions on Vehicular Technologies*, **53** (1), 156–162.

[18] M. Valkama, Advanced I/Q signal processing for wideband receivers. PhD dissertation. Department of Engineering, Tampere University of Technology, Tampere, 2001.

[19] Antttila, L., Valkama, M. and Renfors, M. (2008) Frequency-selective I/Q mismatch calibration of wideband direct-conversion transmitters. *IEEE Transactions on Circuits and Systems II: Express Briefs*, **55** (4), 359–363.

[20] S.A. Bassam, M. Helaoui, S. Boumaiza, and F. M. Ghannouchi, Experimental study of the effects of RF front-end imperfection on the MIMO transmitter performance. Digest IEEE MTT-S International Microwave Symposimum (IMS), Atlanta, GA, June 2008, pp. 1187–1190, 2008.

[21] Bassam, S.A., Helaoui, M. and Ghannouchi, F.M. (2009) Crossover digital predistorter for the compensation of crosstalk and nonlinearity in MIMO transmitters. *IEEE Transactions on Microwave Theory and Techniques*, **57** (5), 1119–1128.

[22] Tse, D. and Viswanath, P. (2005) *Fundamentals of Wireless Communication*, Cambridge University Press, New York.

[23] Sulyman, A.I. and Ibnkahla, M. (2008) Performance of MIMO systems with antenna selection over nonlinear fading channels. *IEEE Journal of Selected Topics in Signal Processing*, **2** (2), 159–170.

[24] L.M. Davis, Scaled and decoupled Cholesky and QR decompositions with application to spherical MIMO detection. Proceedings IEEE Wireless Communications and Nerworking Conference, New Orleans, LA, March 2003, pp. 326–331, 2003.

[25] I. Medvedev, B.A. Bjerke, R. Walton, *et al.* A comparison of MIMO receiver structures for 802.11n WLAN – performance and complexity. Digest IEEE International Symposium on Personnal, Indoor and Mobile Radio Communications, Helsinki, Finland, September 2006, pp. 1–5, 2006.

[26] Proakis, J.G. (2005) *Digital Communications*, McGraw Hill, New York.

[27] Y. Palaskas, A. Ravi, S. Pellerano, B. R. Carlton, M. A. Elmala, R. Bishop, *et al.*, A 5-GHz 108-Mb/s 2x2 MIMO transceiver RFIC with fully integrated 20.5-dBm P1dB power amplifiers in 90-nm CMOS, *IEEE Journal of Solid-State Circuits*, **41**, 12, 2746–2756, 2006.

[28] Kumar, M., Tan, Y. and Sin, J.K.O. (2002) Excellent cross-talk isolation, high-Q inductors, and reduced self-heating in a TFSOI technology for system-on-a-chip applications. *IEEE Transactions on Electron Devices*, **49** (4), 584–589.

[29] C.S. Kim, P. Park, and J. Park, Deep trench guard technology to suppress coupling between inductors in silicon RF ICs. Digest IEEE MTT-S International Microwave Symposium (IMS), Phoenix, AZ, May 2001, pp. 1873–1876, 2001.

[30] Kim, H., Jenkins, K.A. and Xie, Y. (2002) Effective crosstalk isolation through p+ Si substrate with semi-insulating porous Si. *IEEE Electron Device Letters*, **23** (3), 160–162.

[31] Wu, Y.H., Chin, A., Shih, K.H. *et al.* (2000) Fabrication of very high resistivity Si with low loss and cross talk. *IEEE Electron Device Letters*, **21** (9), 442–444.

[32] Saffar, D., Boulejfen, N., Ghannouchi, F.M. *et al.* (2011) Behavioral modeling of MIMO nonlinear systems with multivariable polynomials. *IEEE Transactions on Microwave Theory and Techniques*, **59** (11), 2994–3003.

[33] Eun, C. and Powers, E.J. (1997) A new Volterra predistorter based on the indirect learning architecture. *IEEE Transactions on Signal Processing*, **45** (1), 223–227.

[34] A. Zhu, and T. J. Brazil, An adaptive Volterra predistorter for the linearization of high power amplifiers. Digest IEEE MTT-S International Microwave Symposium (IMS), Seattle, WA, June 2002, pp. 461–464, 2002.

[35] Crespo-Cadenas, C., Reina-Tosina, J., Madero-Ayora, M.J. and Munoz-cruzado, J. (2010) A new approach to pruning Volterra models for power amplifiers. *IEEE Transactions on Signal Processing*, **58** (4), 2113–2120.

[36] Roblin, P., Myoung, S.K., Chaillot, D. *et al.* (2008) Frequency selective predistortion linearization of RF power amplifiers. *IEEE Transactions on Microwave Theory and Techniques*, **56** (1), 65–76.

[37] Bassam, S.A., Helaoui, M. and Ghannouchi, F.M. (2011) 2-D digital predistortion (2-D-DPD) architecture for concurrent dual-band transmitters. *IEEE Transactions on Microwave Theory and Techniques*, **59** (10), 2547–2553.

[38] Bassam, S.A., Chen, W., Helaoui, M. *et al.* (2011) Linearization of concurrent dual-band power amplifier based on 2D-DPD technique. *IEEE Microwave Wireless Component Letters*, **21** (12), 685–687.

[39] R. N. Braithwaite, Digital predistortion of a power amplifier for signals comprising widely spaced carriers. Digest of the 78th ARFTG Microwave Measurement Symposium, Tempe, AZ, December 2011, pp. 1–4, 2011.

[40] Liu, Y., Chen, W., Zhou, J. *et al.* (2013) Digital predistortion for concurrent dual-band transmitters using 2-D modified memory polynomials. *IEEE Transactions on Microwave Theory and Techniques*, **61** (1), 281–290.

[41] Kim, J., Roblin, P., Chaillot, D. and Xie, Z. (2013) A generalized architecture for the frequency-selective digital predistortion linearization technique. *IEEE Transactions on Microwave Theory and Techniques*, **61** (1), 596–605.

[42] Younes, M. and Ghannouchi, F.M. (2013) On the modeling and linearization of a concurrent dual-band transmitter exhibiting nonlinear distortion and hardware impairments. *IEEE Transactions on Circuits and Systems I: Regular Papers*, **60** (11), 3055–3068.

[43] M. Younes, A. Kwan, M. Rawat, and F. M. Ghannouchi, Three-dimensional digital predistorter for concurrent tri-band power amplifier linearization. Proceedings IEEE MTT-S International Microwave Symposium (IMS), Seattle, WA, June 2013, pp. 1–4, 2013.

[44] Younes, M., Kwan, A., Rawat, M. and Ghannouchi, F.M. (2013) Linearization of concurrent tri-band transmitters using 3-D phase-aligned pruned Volterra model. *IEEE Transactions on Microwave Theory and Techniques*, **61** (12), 4569–4578.

[45] M. Younes, Advanced digital signal processing techniques for linearization of multiband transmitters. PhD dissertation. Department of Electrical and Computer Engineering, University of Calgary, Calgary, 2014.

[46] Younes, M., Hammi, O. and Ghannouchi, F.M. (2011) An accurate complexity-reduced 'PLUME' model for the behavioral modeling and digital predistortion of RF power amplifiers. *IEEE Transactions on Industrial Electronics*, **58** (4), 1397–1405.

[47] Mirri, D., Iuculano, G., Filicori, F. *et al.* (2002) A modified Volterra series approach for nonlinear dynamic systems modeling. *IEEE Transactions on Circuits and Systems I: Regular Papers*, **49** (8), 1118–1128.

[48] Safari, N., Rste, T., Fedorenko, P. and Kenny, J.S. (2008) An approximation of Volterra series using delay envelopes, applied to digital predistortion of RF power amplifiers with memory effects. *IEEE Microwave Wireless Component Letters*, **18** (2), 115–117.

[49] Younes, M. and Ghannouchi, F.M. (2015) Behavioral modeling of concurrent dual-band transmitters based on radial-pruned Volterra model. *IEEE Communication Letters*, in press.

Index

Behavioral Modeling and Predistortion of Wideband Wireless Transmitters, First Edition.
Fadhel M. Ghannouchi, Oualid Hammi and Mohamed Helaoui.
© 2015 John Wiley & Sons, Ltd. Published 2015 by John Wiley & Sons, Ltd.